The Secrets of Loving Touch

Franz Benedikter

The Secrets of Loving Touch

How the loving touch releases hormones
that benefit health, happiness, and beauty
and are essential for complete physical
and spiritual well-being

Translated into English by Christine M. Grimm

LOTUS LIGHT
SHANGRI-LA

1st English edition 1996
© by Lotus Light Publications
Box 325, Twin Lakes, WI 53181
The Shangri-La Series is published in cooperation
with Schneelöwe Verlagsberatung, Federal Republic of Germany
© 1990 reserved by Windpferd Verlagsgesellschaft mbH, Aitrang
All rights reserved
Translation by Christine M. Grimm
Layout by Monika Jünemann and
panta rhei!, Mediaservice, Uwe Hiltmann, Niedernhausen/Ts.
Cover design by Wolfgang Jünemann
ISBN 0-941524-90-6

Printed in the USA

4

Table of Contents

Introduction 7

The Secrets of Loving Touch 9
Influencing the Psyche Through the Body 9
The Physical Aspect of the Psyche 11
Adrenaline and Noradrenaline—The Stress Hormones 11
Endorphins—The Mood-Lifters 12
The Vital Hormones—Desire for Love 13
The Partnership Hormones 13
Still More Hormones for Sensuality 14
Constant Stress Tenses the Muscles and Blocks the
Flow of Energy 15
Tactile Hypnosis 17
Falling Asleep Easily 19
Tender Touching 21
Letting Go—The Feeling of Weightlessness 22

Massage—Stroking and Caressing 25
The Pharmacological Laboratory Within Us 30
Awakening the Body 31
Beginning of the Exercise Program 34
The Hands 35
Kneading and Rolling 35
Stroking—The Understanding Touch 38
Hand Contact 40
Finger and Thumb Pressure 43

The Position of the Trigger Points and Zones 45
Trigger Points and Trigger Zones 45
The Skin as Contact Organ 48

Introduction to Endogenous Induction 50
The Restoration of Hormonal Harmony 50

The Three Different Types of Treatment   55
    Freeing the Psyche of Inhibitions and Blocks   56
    Deepening the Contact Between the "I" and the "You"   57
    Feeling the Contact to the Human Environment   59

The Exercises for Endogenous Induction   61
    Switching from Tension to Relaxation   61
    Reducing Adrenaline and Building Up Endorphin   61
    Treatment Summary   62
    The Relaxation Exercises   64

Releasing Inhibitions and Blocks   80
    Exercises Which Can Be Done Alone   80
    Summary of Individual Treatment   80
    Individual Exercises   86

Deepening the Contact Between
the "I" and the "You"   117
    Exercises for Partners and Friends   117
    Summary of Partner Exercises   118
    Partner Exercises   121

Feeling the Contact to the Human Environment   135
    An Exercise That Can Be Done in a Group   135

In Closing: Change is Growth   137
About the Author and His Work   140
Addresses   141
Literature and Music Tips   142

Introduction

The New Age of Tenderness

The Secrets of Loving Touch is an unbelievably effective program in which certain skin zones are touched to release extremely relaxing, mood-lifting or euphoretic hormones which make you balanced, healthy, and happy. In addition, it instructs us on how to free ourselves of inhibitions and blocks and thereby increase the entire strength of our personality.

Through the psyche, a gentle, tender touching of the skin triggers hormonal reactions. Endorphins bring feelings of happiness, increase the willingness to perform, enhance the awareness of vital consciousness and intensify the sensory perception. Estrogen and testosterone are true energy carriers and rejuvenating agents.

Above all, tenderness is skin contact, the feeling of warmth and closeness. People who do not allow these feelings forget them. They atrophy like muscles that are no longer used or trained. Being together with a partner, however, does not necessarily mean being completely satisfied in terms of tenderness. Too many of us have forgotten what tenderness is. On the other hand, "stroking studios," tantra seminars, and tenderness training are evidence of the longing to touch and be touched. Perhaps it is possible to live without tenderness, but it is not possible to be happy at the same time. People who are not touched tenderly and stroked are less open. They become unsociable and lose their warmheartedness.

A healthy person has a great longing to be touched. Sensitive touching has a very relaxing effect and is a sensuous experience at the same time.

Endogenous induction is what we call what we are doing with the exercises introduced in this book. This is what takes place when they are done: the effect which comes from a certain part of the body and its skin zones stimulates the body to achieve a state of hormonal harmony from within itself.

The compact exercise program introduced here shows how, by touching certain trigger zones of the body, we can have a positive effect on the body through the psyche by way of self-massage and partner massage, which is much more like a tender touching. In doing so, we teach our body to make available certain hormones in exactly the amounts we need for a happy, healthy and liberated life.

Endogenous induction assumes that it is not necessary to externally supply hormones which our body can itself produce in an adequate amount. We only have to learn how we can support this self-harmonizing process. *The Secrets of Loving Touch* is a way of creating this soothing equalization. In doing so, our body shows us that it is our best friend.

All of these exercises have been tested in long years of practice—and I would like to thank all the participants for the comprehensive and open reports on the effects which the exercises have evoked within them. Time and again, it has been an unbelievable experience for me that I am permitted to witness how the positive changes in the facial expression, the posture, and in the personal charisma of the participants have been triggered by the exercises. I have often seen how the emotional condition of a person has completely changed through one single exercise, how despair has become security, how pessimism has become optimism and how a state of being caught within oneself has turned into a pleasant openness.

When I look at my work and this book with this perspective, I have the feeling that, if given proper care, a rather small, delicate, and sensitive plant with deep and good roots will become a strong, large tree ... a being that can live both its own strength and weakness at the same time.

I wish all of us the openness for changes and growth on all levels.

The Secrets of Loving Touch

Influencing the Psyche Through the Body

More than we commonly believe, our entire life is regimented by dependencies, by a continuous succession of cause and effect. And our mental, emotional, and sexual activities are largely determined by our physical condition. There is also a causal relationship here.

The way in which we think, what we feel and want, is dependent to a great degree upon our respective hormonal situation. This is dependent upon the amount of hormones respectively produced or secreted, which have greatly differing effects.

Hormones therefore contribute to whether we are happy, healthy, and beautiful.

In turn, our hormonal situation is controlled on the one hand by physical triggers and by psychological ones on the other hand.

The harmonious and enormously complex interplay of all the organs (heart, blood circulation, liver, nervous system, lungs, kidneys, stomach and other organs), the muscle groups and the endocrine system determine our overall well-being.

When our attention is completely concentrated on a painful stomach, because we have eaten the wrong things or too much food, we can no longer think clearly, feel, or live out an instinct. This can also manifest itself as desire, for example, the *desire* to enjoy music; as *need*, for example, the need to peacefully drink some tea; or as *interest*, for example, the interest to read an article or book. The flow of thoughts is replaced by the perception of pain, and feelings and instincts are overshadowed by it.

The interactions of the body and the psyche are inseparable—and this interplay has an effect:

- **On our way of thinking**. The amounts of adrenaline, noradrenline, and dopamine released, as well as the vital hormones testosterone and estrogen, have an effect on our thoughts. Adrenaline accelerates our mental activity. In contrast, the vital hormones influence the life of thought through an entire range of attitudes which can fluctuate between optimism and pessimism, euphoria and depression, an ability to concentrate and distraction, potency and impotency. As explained in the following box, our interest is concentrated on certain hormones which are decisive within a complex system. We will examine the hormonal systems, the effects and counter-effects of which are understandable—those upon which we can have an influence with the following exercises.

- **On our way of actively living and expressing feelings**. Within all people there are fluctuations between going outside of oneself (being extroverted) and drawing back into oneself (being introverted), between inhibitions and disinhibitions, affection and aversion in all conceivable gradations. We can feel free or suffer under the tensing effects of blocks. The "physical" hormonal situation is always decisively involved here as well. In *emotional* people, for example, there is a preponderance of released endorphins which can be observed.

- **On the instinctual life** in the area of intimate, occupational, and socio-cultural interpersonal relationships. Drive, strength, and resoluteness correspond to the amount of the vital hormones testosterone and estrogen which have been released.

As we become aware of this meaning of the body, we can sense the need to concern ourselves more extensively with our body and experience it more intensively. This initially applies to the healthy person who wants to achieve his or her highest level of performance, but also particularly for those who suffer under certain disturbances like weak concentration, a lack of *power* and vitality or inhibitions in the emotional area, and a libido (sexual desire) which is too weak.

The Physical Aspect of the Psyche

Our attitudes, our psychological condition and our activities are continuously influenced by biochemical processes which take place within our organism and represent the other—the organic—aspect of the psyche.

Biochemical processes are the effects of a very large number of substances which are either released or blocked. There is apparently a superior controlling authority for the hormones, the psychological control substances, which is entitled to the final decision. This is the "self," the result of all the perceptions stored in the memory during the course of a lifetime. These perceptions are intuitively consulted and lead to the making of decisions.

Here we will only describe some of the most important hormone systems, those whose mode of action on the life of thinking, feeling and sexuality appears to be certain. We will learn how these hormones are stimulated with the exercises in the second part of the book. At the same time, this is the way to effect the psyche through the body.

Adrenaline and Noradrenaline—The Stress Hormones

Adrenaline and the similar *noradrenaline* have the effect that the body and mind directly prepare to be able to immediately provide their top level of performance in an emergency situation. Seen in terms of the history of evolution, the human being was therefore prepared to show the right reaction necessary for survival in a moment of acute danger when the required response was to attack or flee.

Adrenaline accelerates the pulse rate and breathing rhythm, narrows the blood vessels and thereby increases the blood pressure, intensifies the performance of the organs and the metabolic process. All of this has the effect, in accordance with the personal state of experience, of creating a condition of tension and the highest level of alertness.

Today, the danger situations have shifted extensively—from physical to psychological stress. We mainly find these situations in partnerships, in the family, in the social area and particularly in working life—areas in which we must achieve the highest levels of performance. The achievement of a high level of performance becomes a permanent condition for many people. This means that we, consciously or unconsciously, continually put ourselves in a psychosomatic state of tension which does not correspond to our actual nature and often also exceeds the limits of what is tolerable: constant stress. This means that we are overstraining our neurons, our information network, putting ourselves into a *neuro*tic condition in which what is actually human (such as a large portion of the feelings, of which only the feeling of fear survives) can no longer be lived.

Endorphins—The Mood-Lifters

Endorphins are morphine-like substances supplied by the body under certain conditions. They create well-being and good feelings. They used to be the body's own pain-killing agents, but today, where our body no longer differentiates between physical and psychological pain, their absence also means physical pain and generally signals a lack of well-being.

This is not so much a matter of good sexual feelings, but the psychophysical feeling of warmth and security that we perceive in intimate, emotional relations. The endorphins appear to be particularly linked to a major part of the mother-child relationship. In a certain sense, they are the antagonists of adrenaline, producing relaxation and guiding us to human closeness.

The Vital Hormones—They Create a Desire for Love

Testosterone/Estrogen: These hormones, substances which are produced in the testicles and ovaries, have a close correlation with the sexual activities and physical vitality. They effect the various phases of reproduction, which includes sexual desire, sexual intercourse, pregnancy and birth. They frequently create an inner feeling of sensuality, stimulation, strength and also the drive to measure oneself against the world around us (also in terms of *mental performance*) and have a large influence upon vitality.

Hormones which are still largely unknown in terms of their composition, though we clearly experience their effects because they regulate the physical relationship to the partner, belong to the fourth system. These are the partnership hormones.

The Partnership Hormones

We can consider these hormones to be the biochemical ambassadors between one body and another body—with an apparent reciprocal action on the psyche. They are transmitted through the skin and perceived as pheromones (odorous substances) through the sense of smell or the sense of taste.

Andosterone is produced by those glands with an outward orientation, particularly in certain zones: chest, armpits, and in the genital area. Andosterone is also present in the saliva: it generates and increases sexual desire when there is mouth contact—when kissing.

Skin substances are produced by the skin glands, particularly those of the lips and the gums. We know this much about their effect: they heighten the desire for intimate and erotic contact and trigger general excitation. There are naturally also possible anti-effects: a smell which is strong or too strong can naturally also have an inhibiting effect. Personal receptivity is the deciding factor. Those who have an aversion to certain smells block the natural effect of these odorous substances and can even produce hormones which have an antagonistic effect. A shower before an encounter must be followed by a subsequent secretion of sweat in order to offer enough olfactory sensory stimulus for people don't get aroused very quickly: exuberant dancing or gymnastics which have a strong warming effect can help here. Cosmetics offer many types of deceptions for this purpose: lipstick simulates a stronger blood circulation and certain perfumes try to replace the effect of the hormones normally contained in perspiration. However, we can create everything that these means offer us through our own hormones.

Among the partnership hormones are also the accompanying effects of *dopamine* and *noradrenaline*. Both of these are actually neuromodulators and influence the life of thought. They are produced in the brain and in addition to their main task, the sensation of pleasure, they create greatly differing types of desires such as hunger, thirst and sexual drive. They primarily react to visual-sense stimulus and, in a much lesser degree, to smell and taste.

Still More Hormones for Sensuality

It is also interesting to know what has been assumed up to now about hormones which cannot be consciously influenced.

Oxytocin is a brain hormone which causes the involuntary contraction of certain muscles. In the woman, this mainly means the contracting of the uterus during birth and the flowing of breast milk. The effects on the sexual organs are: the stimulating contraction of the penile muscles and vaginal muscles—contractions which can also be consciously controlled. Oxytocin has a further remarkable psychological effect: it stimulates dreams of love and the desire for closeness and tenderness.

Ayetylcolin is found in the nerve fibers of the clitoris and the penis. It increases the blood supply and causes the erection of the penis and the clitoris.

Luliberin exhibits its effect in the hypothalamus and has a controlling function for the vital hormones testosterone and estrogen. Desire for pleasure is increased in this manner.

Prolactin exists in both men and women. It inhibits the desire for pleasure and causes—in larger amounts—impotence in the man and orgasm problems in the woman.

Progesterone is created in the ovaries and released during pregnancy. It is similar to estrogen and has—in reciprocal action with it—effects on the sexual life.

These are the currently valid endocrinological views of the psychosomatic effects of the hormonal systems which are most important for us, in so far as they are of interest for us in this book.

Constant Stress Tenses the Muscles and Blocks the Flow of Energy

Stress creates tension—and it always does this when we are subject to stress-promoting situations for a longer period of time. On the physical level, we force ourselves to an increase of activity which is actually no longer physically tolerable.

This can occur consciously, such as when we have an exaggerated urge or compulsion to achieve, or unconsciously because of a general feeling of fear. On the **psychological level**, we can perceive the tensions are created because we become nervous, tense and on edge. But it is simpler and more apparent to feel them as **physical tension**.

Individual hardened muscle groups, such as those in the stomach area, at the back of the neck, or in the chest area, let this tension become just as painful as it is harmful: sooner or later, but very

15

definitely at some point, irreversible damage will occur to the various organs in that their function will initially be disturbed: This is then followed by chronic organic diseases.

It becomes dangerous for our health when we have gotten accustomed to this pain and no longer perceive the signals that the body wants to convey to us through the pain.

When we are in this state and feel our body with a probing, sensitive hand we can determine the hardening in a simple type of self-diagnosis: certain muscles are hard, tensed, and also energetically blocked in extreme cases. If we now simply "massage" them, we will first feel a slight pain, but the warm hand can release the tension after a short time and thereby create the wonderful feeling of well-being that basically goes along with relaxation—always when we more or less intensively "switch" from tension to relaxation.

Who has never, particularly when in a stress situation, longed for the good feeling that a simple walk on the beach or rambling through green meadows can create? In moments when one's own fears become overwhelming, who has never wished for a friendly arm around the shoulder or some other expression of love, of being understood and protected?

When we feel this wish, then we *must* satisfy it because we can satisfy it—also within our own limited possibilities which always offer us more than we can imagine at first. What stops us from satisfying our desires is usually just the inhibitions, which means the greatly varying types of *psychological* resistance. Yet, we can have an influence upon them.

In concrete terms, this simply means: Get up from your chair, put down your "tools," start with two steps in order to slowly turn your thoughts away from what is preying on your mind. With this small act of will, we open the door to a new world which we can experience as a subtle dream experience or as an intensive act of imagination.

Tactile Hypnosis

When we are aware that we can effect our well-being in very different ways, why do we not simply "rise up" and bring our body into a pleasant, relaxing position?

1. Because the initial tension is still too great: to lie down, to give up the activity, means that we put ourselves fully at the mercy of the tension because we give it space when we interrupt the activity—which has nevertheless been diverting us. We must understand this. We let go and surrender to the falling phase, but it is soon followed by the rising phase. The perception of fears is replaced by a pleasant feeling.
We simply have to stroke ourselves and feel our warming hands on the painful places. In this way, we initiate a simple hypnosis through the sense of touch. We abruptly replace the vision usually used in hypnosis through touching, which is sensorially more perceptible and can completely take hold of us, letting us forget our troubles—simply hypnotized.

2. Because we have unlearned how to be like children—children whose instincts are still unadulterated and not deformed, who simple start to run, tussle and play, and move their bodies without subjecting themselves to imposed limitations.
The *adult human being* is no longer even permitted to cry, freely express his feelings, move the body as he likes because it is not proper and he has neither time to do so—nor, more precisely, does he take the time.

3. We ultimately no longer know what is good and pleasant for us: We are simply "over-sophisticated," have an extremely large amount of technical—in social terms—*useful* knowledge, but we have far removed ourselves from understanding the soul, psyche and body with their appropriate demands. We have forgotten how to speak with our organs and to listen to what the liver, stomach, lungs, or particularly the *heart* can say to us. We usually only

listen when our organs give such intensely painful signals (illness) that we can no longer ignore them.

But at this point in time, body and soul have often already experienced an extreme loss in the quality of life—they vegetate in a type of dark tunnel, which requires some effort to get out of again. And this only happens when we once again shift the priorities.

Our well-being is also an important task—and not just with respect to ourselves. If we cannot or do not want to give everything that society and family demand of us, then we just have to learn to concentrate on things which are more important for our philosophy of life. But very few people do this. We basically no longer *want* to understand the needs of our body. We neglect it, we abuse it by exploiting it, we mistreat it mercilessly without knowing whether we will later be capable of paying the bill that it will present us.

The bill will be presented to us in any case—if not today, then tomorrow. We will reap what we have sown—our philosophy of life creates the contents of our life.

If someone suffers from headaches, he is showing an "alarm signal" which should be taken very seriously since it signals an overstrain of the mind that is not harmless. This type of extreme mental fatigue can occur consciously or subconsciously (then we do not "notice" the exploitation of the understandably very pain-sensitive brain cells). Untouched by this, the person carries on more or less indifferently in this manner and lets his organ, which has given this type of a clear sign of overfatigue, continue to work at full speed. Perhaps he helps himself with a pain-killing tablet, strong coffee or biochemically by releasing even more adrenaline and continuing to "whip himself" into the highest level of performance.

He does not listen to this warning as long as there is still time for it: by resting, relaxing, or doing physical exercise he could have the blood flow back out of the brain and into the body, thereby achieving a reduction of brain activity.

Too seldom do we look for the causes of pain. We satisfy ourselves with fighting the symptoms at times and only react to the pain threshold. As long as this functions and is still tolerable, we act like the driver of a car who continues to put his foot on the gas pedal when the engine is already making frightening sounds.

A person who suffers from sleeplessness because he is overtired when he goes to bed with too many problems raging in his brain falls asleep because of physical tiredness without having straightened up things in his head, even though he knows that his overly intensive thought activity in the form of dreams will chase him out of sleep much too soon ("I simply can't go back to sleep!"). This repeats itself evening after evening. The person might help himself with tablets which only subdue the problems or postpone their perception for eight hours at most.

Falling Asleep Easily

It would be very good for us to be concerned with ourselves a bit before going to sleep and not simply fall worn out into bed. But we must also first learn to do this, just like we have to learn to control our energy household.

Just as obviously as we brush our *teeth* in the evening, we should also provide clarity for our minds and put our thoughts in order. When we are together with a partner, we should talk to him or her—about everything that is important to us. If we are alone, it helps to go out on the balcony or open the window and consciously breathe in the fresh night air or have a look at the stars—and take in the peace that the sight of the eternal cosmic order awakens within us. If we simply let ourselves have the time, the self-regulating mechanism of our mind begins to work on its own, creating order and turning itself off.

How we can learn to once again give this positive-acting mechanism space is described on the following pages on the basis of very simple exercises. They are the best help for helping yourself.

Paradoxically, it sometimes appears to be the hardest thing to find out what is good for ourselves. Even if we are sensitive enough to perceive the imbalances within our own energy system, we sometimes do not know how we can translate these perceptions into action.

Difficulties in going to sleep are enormously wide-spread today. But a sleeping pill can hardly do more than what a glass of wine can do. Both are actually drugs and have their side-effects. What we cannot erase in our overloaded memory with the conscious use of a *little* alcohol, cannot be erased by the best tablet either.

Many of our natural bodily functions and muscle activities are limited because we sit at our jobs, while driving and working at computers. There is basically little that we can change about this since it is a component of our life today and we are subject to certain social standards in this respect. Yet, at the same time, we should not forget that not only the economic sustenance of life but also one's own physical well-being is an absolutely essential value and that we can positively link these two apparently antagonistic necessities with each other. Performance and physical and emotional health do not have to exclude each other if we integrate both of them into a healthy rhythm. After a week of constant stress, it is much more difficult to reduce the tension than when this is done once a day. It is precisely our overemphasis on the mind, which has distanced us from a natural way of handling our body, that can help us by becoming conscious of taking this step in the reverse direction once again, coming back into contact with our actual nature and allowing the body its right to good feelings.

Both our muscles and our organs will thank us for the attention given to them by providing us with a new feeling of well-being. This is our very first goal in the following exercise program. In the long run, we want to maintain our vitality, youthful vigor, and a fit organism that gives us good feelings.

In principle, the "massage" introduced on the following pages is based on the way loving parents touch or treat their children.

Tender Touching

We experience the most basic form of tender touching in earliest childhood. We watch a mother who lovingly goes to the cradle, to her crying child. She touches it in a very distinct way and cradles it in her arms. Then she rocks and strokes it, pressing her own cheek to the baby's cheek; she puts her warm hand on its little stomach, which already shows the first signs of the "tension" which much later and much too frequently torments our abdomen. The baby's bottom is patted—in a very distinct rhythm. With the child in her arms, the mother moves and rocks it even in the way she walks. Time and again, a "wonder" happens here—crying turns into a smile.

How a mother and child touch

If someone then says that the child is being spoiled, they lack in understanding and are being heartless. The little person often cries "just" because it must painfully learn to separate from the totality. Giving love and receiving love is always an inevitable precondition for a psychologically healthy life. Both are just as necessary in the long term as food and drink. Just as a person can die of thirst because of a lack of water, he or she can also dry out because of a lack of love.

Seen in psychosomatic terms, one could say that children suffer and cry when they have an endorphin deficiency. This deficiency can be eliminated by patting and cradling, through skin contact and rhythmic movements—when this is done, this "endogenous analgesic" is once again more intensely released and the child experiences feelings of happiness.

Letting Go—The Feeling of Weightlessness

Another important experience is the feeling of unconditional trust, being able to let oneself fall. The father lets the baby feel the weightlessness, tosses it a bit into the air in order to soon catch it again with secure hands. The child's expression is quite meaningful: at first it is fearful, then overjoyed. In the process, the father did not do anything other than call to mind the weightlessness of which the baby still has a strong recollection, as it was experienced within the womb—the earthly paradise of human beings—up until the time of birth.

Which of us have certainly at least once not felt the desire—even if it wasn't completely in this manner—to be treated with at least a similar attitude by his or her partner or friend? Why are we sometimes ashamed to show this elementary need? We have not only an inner child; there *is* also a *child* with all its needs still within us and this is what it always will remain. This weightlessness is part of the ritual of carrying the bride over the threshold. In a loving sexual exchange, this weightlessness is also a part of "flying"!

The father lets his child feel the weightlessness

Loving touching—like parents do with their children—has the effect of releasing endorphins, when seen with a purely physiological and endocrinological perspective: the natural antidote for physical as well as emotional pain. It is the instinctive act of someone who only wants the best for a beloved being.

If a child cries because it has stomach pain, a physiological reasons, the crying is understand as a cry for help and a doctor is called. But if the child cries because of a need for love, for psychological reasons, this is not understood as a cry for help. However, this is what it is and attention should be paid to it just as much, if not more, since psychological pain can often be more painful than physical pain.

As adults, we have frequently not only forgotten how to cry, but we have also forgotten the other possibilities of calling for help when we are sad, our "heart" somehow hurts, or because we cannot receive or give enough love.

Tolerating a psychological pain for a longer period of time—and even if the cause of it appears totally unchangeable to us—is doing violence to our soul. If our shoes are too tight, we take them off. It is easier for us to perceive and treat physical pain than our own psychological pain. It is here that we are all too willing to tolerate and suppress things. Suffering is normal today, and we *suffer without crying*. The pain is no longer recognized by either ourselves or by other people, and we have sometimes suppressed it to the point of no longer being able to perceive it.

We know that our ability to achieve sinks when we are sad! For this reason alone, we should recognize "being sad" as an expression of psychological pain and treat it: otherwise, in the end there is all too often the depressed person who stands in the darkness of his uncried tears, swallows his emotional pain, and suffers and vegetates more than he lives.

Massage—
Stroking and Caressing

Muscles which are tense and hardened or even blocked for a longer period of time lead to a part of the blood vessels being narrowed; the blood can no longer flow optimally, and the free flow of life energy is limited or at least strongly hindered to the disadvantage of the entire body. *Massages* are very helpful in this case. Cramped muscle zones are "loosened"—and the entire body relaxed under the gentle pressure of practiced hands. The flow of blood, which now flows more freely, creates a feeling of well-being. Unfortunately, this usually doesn't last for very long.

The tension is soon reflected in the body once again: the psyche becomes somatic. The massage can only have an effect on the symptom (the muscle tension), but it cannot influence the deeper cause which is the "tension" caused by a certain psychological condition.

If we compare the effect of a classic massage with a simple caress or an understanding embrace by the partner, family member, or an "intimate" friend, the difference becomes apparent. The effect of this commonplace and emotional gesture is by far more intense than the most professional treatment by a "masseur."

The following instructions for "massaging" the body are therefore not instructions for any type of massage in the usual sense.

The exercises simply concentrate on creating a pleasant and very effective connection between the hand and the body. This is why we are not talking about a "technique," but instead about "understanding touching": we are touched with understanding by our own hand or the hand of a loving partner who is perhaps unexperienced in massage.

When we, for example, massage the soles of the feet in this manner it is not to create a connection to a specific organ and have an effect on it, but rather only to *experience* a sensation—something like a gentle "jolt" of energy which reaches the stomach *area* from the sole of the foot. This permits us to sensitize our body in a gentle way and achieve a more intensive relationship to it. This pleasant sensation triggers a "tactile hypnosis," which means nothing more than switching from tension-promoting thought to relaxing feeling. "Just" feeling instead of thinking for half an hour creates a state of happiness, a feeling of pure well-being which we often believe we have lost.

Very simple stroking effects hormonal changes in the body, which means changes in the endocrine system as well as in the central nervous system. The length of this loving treatment is naturally decisive for the lasting and far-reaching hormonal effect.

We like to stroke: when a cuddly, soft-furred creature comes to us, we automatically reach out our hand to enjoy the warmth and softness. And when it is a cat, it will soon respond with a pleasurable purring and show us that petting is exactly what it wants

In simplified terms, this has the following significance for a partnership: as long as a "couple" strokes, the love lasts. When they no longer stroke each other, the love also comes to an end. The marriage or partnership turns into just living together. Stroking—like every form of skin contact—is a very direct and effective expression of "love," the physical basis of this phenomenon which appears to be so emotional to us.

By accepting the interaction of cause (love) and effect (stroking), it becomes clear to us that we can have a conscious influence on the flow of energy by also being *loving* to each other when we are tired, need rest, and feel exhausted. This deliberate trigger releases much positive energy: If I stroke someone with understanding, his or her eyes will soon beam at me—it is giving and taking. And this is more than just a little thing.

As we have seen in the mother-child relationship, touching the cheeks releases endorphins. The constantly lurking emotional pain is soothed and a feeling of well-being spreads throughout the entire body. This "intimate treatment," which we should have experienced in early childhood, has the effect that we are capable of having very deep emotional experiences in later life as well, even if we just

Stroking results in well-being: the cat purrs

stroke ourselves or—and this is very effective—imagine that we are doing it.

There is no reason to limit this loving togetherness to just love partners (man/woman, mother/child ...). We can also consciously use the loving touch with ourselves (this is described in the first section of the exercises) and with partners to whom we have a trusting and intimate relationship (by *intimate* we mean *human closeness*).

When doing this, it is tremendously important to also open up to the *true experience* since people who only devote themselves to their own imagination remain alone emotionally, isolated behind a glass wall of their own fears and inhibitions. Yet: This separating wall is much easier to overcome than we can imagine.

Simply standing up consciously, leaving the chair in order to take two steps or make a movement that is beneficial for the body (having a good stretch, taking a deep breath) already brings back a bit of the physical feeling—with all of the positive effects described. The oppressed psychological state is then relaxed and sometimes even disappears. It is just as easy to change a rigid attitude through mental impulses, for example, by saying a "good" word to ourselves.

When we do this, we create within ourselves the atmosphere for a friendly contact which immediately changes our facial expression—we can see this by standing in front of a mirror. The physical expression of our face, particularly the synchronization of our eyes and mouth always corresponds to our inner attitude. We once again sense ourselves to be part of our environment and the people who surround us. We feel how much spontaneous willingness there is around us and how the momentary *embarrassment* that we so much fear is just temporary and easy to overcome.

Loneliness is not caused by our character and does not depend on a lack of luck, but is instead a result of our inner attitude, our will. Changing one's own attitude toward other people is an ability that can be learned.

We can change our physical and emotional well-being even through just our *power of imagination*. For example, this correlation has been made scientifically provable by Professor Johann Heinrich Schultz with the therapeutic application of hypnosis in autogenous training.

Try it out yourself: Hold both hands with the palms facing upwards. Then simply concentrate on your left hand or imagine that it is becoming warm. This is exactly what you will feel—namely, that your left hand becomes warmer.

The idea of one's own "warm hand" is actually followed by a measurable increase of the respective skin temperature of up to 4°C and a corresponding relaxation!

Almost a century has passed since the discovery of this reciprocal action. Today we know that just through the visualization of an organ, an "increased flow of blood" will occur to precisely this organ. We can also perceive this energy flow as a feeling of "heaviness." This knowledge is applicable in practical terms for one's own self-healing processes. A woman who is suffering from an inflammation of the ovaries can, in addition to an allopathic or homeopathic treatment, promote the self-healing process through visualization.

The following box describes the scientific view of how we can stimulate our body's own "pharmacology," in as far as we desire this. In many cases, this can be just as effective or more effective than the industrially produced products, medications which we can take for the same purpose.

The Pharmacological Laboratory Within Us

Our Body as a Highly Intelligent Network— A Scientific Excursion

Our organism is a refined pharmacological laboratory that continually produces a very great amount of substances which regulate, among other things, the functions of our brain and our nervous system.

The products which are artificially manufactured by human hands (medications, etc.) which we normally use are nothing other than imperfect imitations of those which are produced by the body. We are only now once again becoming aware of this fact—we just have to think of the trend towards homeopathy.

This fact suggests the thought—as noticed by the experienced scientist Professor Dr. Paolo Pancheri of the Psychiatric Clinic in Rome—that as soon as our knowledge about the natural mechanisms of the organism has been increased, we can find the way in which every person is again capable of newly learning how to set these processes in motion in his or her own body. In other words: We can learn the symbolic language of illness and trigger positive feed-back on the physical level through emotional-mental learning processes.

The suggested path of hormonal change through endogenous induction is the first step in this direction.

Awakening the Body

In order to improve our well-being directly and in a manner which is applicable on a practical level, our nature provides us with various possibilities. This is why there is no harm in using a psychosomatic "trick" now and then.

When we are caught up in our everyday thoughts, which are unfortunately not always pleasant, we can pull ourselves out of this block and feel ourselves "whole" by simply pinching ourselves somewhere in order to provoke a physical experience which shifts our perception.

Each of us has experienced this more than once. When a person who is fearful or suffering from a neurosis because of being tormented by the one-sided overactivity of his thoughts, he can very simply "pinch" himself back into reality (pinch the arm, for example) or drink a glass of water. If someone puts a comforting arm around his shoulder, this can be just as helpful because through the body the access to another, more pleasant-feeling "world" is opened up. It is also very effective to do a half-hour of intensive gymnastics. The way back to physical perception frees the mind which is not meant to function like a computer, the current symbol for pure mental work. Turning off and starting anew sometimes helps get through many of the problems which we get stuck on.

And when wild, painful dreams pursue us during the night and we very intensely experience horrible, unpleasant situations from which we want to flee by any means because they so inconsiderably exceed the pain threshold, there is once again a very simple remedy available to us: create a perception which is related to reality. We can find the way to reality, which is always less horrible, by breathing consciously and deeply or stretching until we wake up. When the nightmare has been overcome, then we can find the way to a pleasant doze until we sink back into deep sleep.

A small boy whose even younger sister was shaken by a hysterical crying fit, instinctively reached for a glass of water and threw it in her face. Without recognizing the cause, he used it to end her despair. The therapy is admittedly rather brutal, but it was nevertheless effective. It does not create a state of well-being, but instead effects a more tolerable state of anger by switching to another level.

Which of us would not rather prefer to be awakened by a glass of water than have to withstand the psychologically painful situation of a nightmare?

It is apparent: When our thoughts torment us, the physical reality comforts us. Our body is also our next and best friend here, a friend we can always ask for help in order to soothe the despair.

At this point, it is very important to me to point out that we see no essential difference between dream and reality (Sigmund Freud also considered "dreaming" to be a trial action).

The poet Wang said: *"Wang dreamed that he was a butterfly, flying between flowers and branches, letting himself be rocked by the wind. Then he woke up and no longer knew if he was actually Wang who dreamed that he was a butterfly or a butterfly which dreamed it was Wang."*

A dream like this—an expression of emotional cheerfulness—is not a gift of God: everyone can have a positive effect on what happens in his or her dreams. Despite this, most of us have too rarely experienced a "pleasant" dream. In a way, it is a law of dreaming that the course of the dream is determined by our basic mood or attitude, which we can influence not only through our imagination. If while dreaming—which means when among the countless thoughts that control the nightly activity of our brain from the subconscious—we associate the vision of something like a *"house"* in a **positive attitude** with *"garden"* and *"picking flowers"* other associations can be connected with the same *"house"*. If the basic **attitude** is **negative**, fearful images result: we see that the *"house burns"* or the "car parked in front of it that won't start at all even though we urgently have to leave." No one comes and our cries for help remain unheard. And *"it begins to rain, the earth becomes soggy, and we are going down ..."* The same fear that determines the activity in the dream also has us suffering during the day.

Wang and the butterfly: the human being between imagination and reality.

In addition to the momentary attitude, there is a further factor which effects the course of the dream. We have determined that what happens in the dream requires a certain intensity so that we become conscious of it and remember it.

In fearful people, *emotional* contents have a stronger value because logic and realism "sleep." These contents are what breaks through the ceiling of consciousness—not "harmless" or calming thoughts.

We have dealt with dreaming here because the following visualization is closely related to the dream imagination.

If it is difficult for us to have an effect on the nightly subconscious course of thoughts, it is therefore usually easier for us to control our imagination during the day.

The exercises of endogenous induction are based precisely upon associating our imagination with perceptions and connecting them with an appropriate treatment of certain points or zones in our body. In addition, the effect of endogenous induction is increased because through the exercises we can particularly effect the essential hormones required for well-being.

Beginning of the Exercise Program

We begin the exercises with a light limbering-up program for body, mind and soul by running in place, doing simple gymnastics and, if we like, dancing to rhythmic music. This relaxes the muscles: the heart beat is accelerated, the breathing becomes more intensive and the body better supplied with blood.

After we have prepared our exercise room (subdued light, telephone receiver off the hook, family members or other occupants have been informed), we consciously dress ourselves: soft cotton, wool, or silk material, close-fitting but not too tight. At least in summer or when the temperature is warm enough, putting on these clothes is more of an undressing than getting dressed. While doing this, we take off annoying pieces of clothing with the concept that we are taking off our problems at the same time.

An affirmation spoken silently or out loud starts the exercises: "Now I am dedicating time to myself; afterwards, I will face the problems of everyday life full of strength and energy."

Now we can begin and more or less follow a ritual to which we dedicate our entire attention. The more intensively we prepare ourselves, the greater the chances for success will be.

With the "relaxation exercises" described starting on page 64, we will relax many of the accessible muscle groups to the point that we are psychologically much more willing to go deeper as well as a result.

At the same time, it is our goal to achieve a complete balance between tension and relaxation in the muscle area.

Before we begin with the actual "massage," we prepare ourselves emotionally.

We put our body into the "balanced position," as described on page 72 and open ourselves to the flow of perception between mind and body. We feel this relationship between mind and body even more intensely after the relaxation exercises. In doing so, our heart beats more strongly, our breathing is faster and more intensive, and we are mentally open enough to deal with the new experience of our physical condition: The body remains passive in the process, the intellect observes and perceives every change.

The Hands

Now we look at our hands, opening them to the flow of life energy which streams from them. They become soft and understanding, as if we wanted to stroke a child. We can naturally use a bit of strength now and then in a loving way.

Only now do we begin with our "massage." There are various techniques available for this purpose.

Kneading and Rolling

We treat the muscle groups (buttocks, neck, back, and nape muscles; abdominal muscles with somewhat more caution), which are not relaxed enough (especially deeper down in the body) and, where stronger blood circulation appears to be necessary, with strong pressure that can reach the pain threshold. Even if this may be somewhat unpleasant at first—it is a symptom that tells us we are at the right place—a pleasant feeling of well-being soon follows.

The possibilities of a self-massage are naturally limited for physiological reasons: We cannot reach every spot on our body ourselves. We pay particular attention to the buttock muscles, where we can massage very intensely. As for the abdominal muscles, this attention applies particularly to women since their bodies frequently tend to take on a "protective" cushion of fat which has more of a

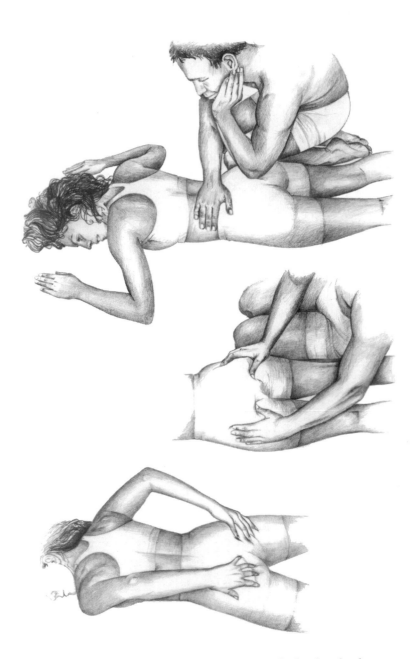

Hands, one's own and those of the partner, which relax the deeper muscles (here: the buttock muscles) with a strong motion.

defensive function than a protective one. This often has psychological causes.

We can understand this when we equate the buttock muscles to a living sensuality and include the abdominal muscles within the area surrounding the ovaries. These falsely appear to lose their sense and function for the woman in a certain psychological situation (disturbed relationship to everything "sexual") and particularly at a certain age (when the "sexual" aspect is absurdly reduced). Those who want to do without sensuality and sexual exchange, or must do without it, should not forget the emotional and *vital* values of these organs and the hormones which are released there. It has been proved that the body ages less chronologically with these hormones than when a conscious or unconscious lack of them has been created. A woman can remain a woman in every respect, even after menopause, even after the removal of the ovaries, because the estrogen production which is important for the maintenance of distinct femininity can once again be stimulated and stabilized with the appropriate exercises. The body can produce estrogen even without the ovaries.

Particularly the abdominal muscles are too tense in many women. For example, feelings of fear can trigger stomach pain. This causes the flow of blood to be blocked and can have unpleasant consequences like a chronically hard abdomen. If this is the case, we use the more gentle and yet effective contact of the laying on of hands as a substitute for the "kneading."

"Kneading" is a strong massage, similar to kneading dough. Decisive for the amount of strength we use to do this are always the perceptions coming from the zone treated. It is naturally possible that a little painful cry can occur, which we should not interpret as a signal to stop. We should imagine that we are in part touching energy blocks and that once again causing this energy to flow may at first be experienced as unpleasant. However, an understanding, strong treatment is the best thing we can do to release these blocks. After the initial expressions of pain, we can stop, wait, and leave space for relaxation. Then we can try it a second time. The initial pain is usually followed by a feeling of warmth because of the more intensive blood circulation and a general feeling of well-being.

We very cautiously treat the spine and the muscles which enclose it. We only use our fingers here, with which we gently roll over the skin alongside the vertebrae. The most important thing is not so much the technical skills, which we can expand with good massage handbooks , but rather the proper understanding of what we are doing: the hands learn to understand the language of the muscles, accept their message, and obey it.

We have thereby created a simple, but intensive physical contact. We also continue to perceive it when the treatment has already ended. By feeling the "forgotten zones" of our body, we also change our attitude towards them.

Stroking—
The Understanding Touch

Now we proceed much more intuitively. We stroke the skin with one or two fingers and change the pressure completely according to the feeling. In doing so, we will be astonished to discover that a stroking which hardly touches the skin can be much more effective that a stronger action. Basically it is a positive thing to change back and forth between the two variations.

In order to determine the effects of finger pressure, after having relaxed a muscle group through hard kneading we can stroke over the zone which we have just worked on with an extremely sensitive touch. This perception is generally very intensive and pleasant.

While doing this, we should pay particular attention to the primary erogenous zones, such as the breast and nipples (in both woman and man—those of a man can also be very sensitive) and the inner thigh.

A twenty-three-year-old woman with two children was advised to do this treatment with her partner, who was still rather immature. However, she could not use it because she knew that her husband

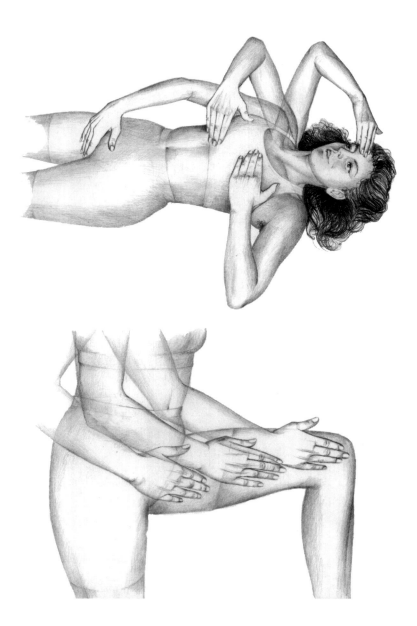

The stroking hand

would not bear the stimulation and that he would even flee from the marriage bed. Nevertheless, this shows that the two people did not have an adequate relationship of personal trust in the physical relationship.

When partners are not familiar enough with the other person's body or do not understand it, they are not adequately prepared for intimate contact. In such a case, the treatment of the intimate zones would also be difficult.

Fundamentally, the treatment of the intimate zones is difficult sometimes. There are, for example, men who flee or cry out when their cheeks are stroked, even if they do not dismiss the whole thing as silly or childish. The sexual contact can then only be limited to the actual sexual zones. There can be no true intimate exchange of energy which equally includes the body *and* mind (psyche), which means the sexual drives *and* feelings. It is clear that this can only result in a purely physical orgasm which makes a person increasingly dissatisfied in the long run. An orgasm which includes mind and body is not possible in this situation.

Many marriages suffer or break up precisely because of this deficit. The purely physical sexual intercourse soon loses its appeal and attempts made with new techniques are unsuccessful. Sexual intercourse is simply misunderstood here, namely in literal terms. It does not create a truly intimate exchange, an encounter, but instead remains just a mutual erotic-motion exercise.

Particularly in self-treatment it is important, if not absolutely essential, to work with creative visualization while stroking. We can remember childhood experiences, the hand of the mother, our first love, or we can dream visions of a fulfilled relationship in the future.

The Hand Contact

A further form of energetic massage is the simple laying on of the hand on certain body zones. To do this, we simply perceive the warmth that occurs in the process. It initially emanates from the hand and then flows into all the body parts which are treated. At the same time, the hand often appears to merge with the skin beneath it

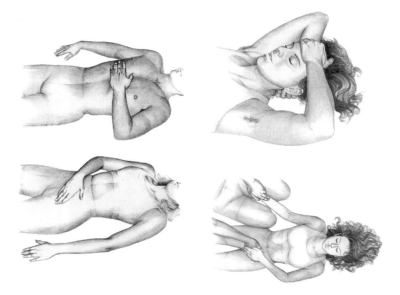

Laying the hand on important zones of the body

or sink into the body. This impression totally corresponds to the actual depth effect of this special treatment.

The impression can also be created that the hand disappears. We no longer feel it and perceive only the warmth of the treated zones. In this case, we have already perfectly attained the desired state of tactile hypnosis (through touch).

Here as well, we can proceed using "pressure" of various intensities and select among a palette of possibilities which range from the most pleasant type to the most effective, whereby the one does not exclude the other. When someone is very sensitive, it can be most effective when the hand hardly touches the skin and almost only the energetic radiation can be perceived.

This effect naturally does not take place immediately. It requires time and the appropriate concentration. Our entire attention should be limited to this point or this zone. If distracting "thoughts" occur, we should not hold onto them, but simply let them go. We ask them to go so that we can deal with them at a later time. We must let go of any type of annoyance or conscious control because these promote tension.

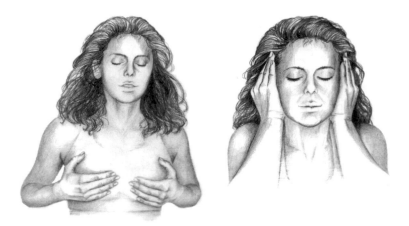

The hand formed like a shell on the chest and on the ear

As an alternative, we can also hold the hand in the form of a cup. We should do this for two places in particular: when we put it on the chest or over our ears we attain a "shell effect."

None of the techniques which have been described up to now should be considered dogma. We solely want to dedicate our attention to the contact of the body and the hand, sensitize the body, and then, and this is most important, promote our own creativity and find new, individually effective treatments. *Every* body has its own very individual demands—people differ very much from each other in this respect. It is sometimes helpful to write down our experiences. On the basis of this "log book," we can more consciously understand changes in our sensual perception, adapt the treatments, and learn many new things.

In closing, just a few words about Wang's dream. Even in the state of tactile hypnosis, in which we have put ourselves through concentration on the body, we can give free reign to our world of dreams and imagination and experience many pleasant things. These trance journeys take much less energy than goal-oriented thinking and are therefore correspondingly relaxing. A difference hardly exists between the reality and the dream reality in this temporary trance state.

How long should these exercises last? Sometimes a few minutes are enough. We do an exercise which we have already tried one more time just briefly or, if time allows, we also dedicate up to a half-hour to our well-being.

After the treatment it is very pleasant to fall into a light sleep. We do not always have to actively achieve something. It is just as useful to prepare our body for achievement.

The Finger and Thumb Pressure

The following description is very helpful if you have had little experience with body work up to now. It also offers itself as self-help in order to find our way back to physical consciousness and once again fully experience our corporeality.

Very intensive perceptions are created when instead of the hand we just use one or two fingers and press on the "specific" point with them. When doing this, we should consider that it is somewhat more difficult with the fingers than with the hand to find the right points such as the solar-plexus or the ovaries.

But we can easily help ourselves at the start: we put the entire hand on the respective zone until the solar-plexus point "announces" itself by radiating heat. Then we know its position and can head for it directly the next time. The effect of finger pressure is usually stronger. Sometimes there is even a feeling that the finger is "penetrating" into us, that it is going into the depths. Other points are easier to reach. The most important points are portrayed in the illustrations on the following pages:

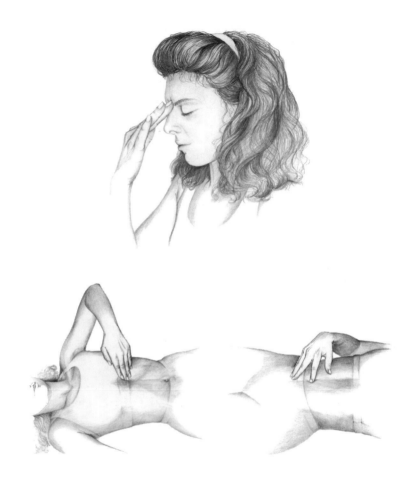

Places where we can exercise strong or gentle finger pressure: the root of the nose (forehead chakra), the navel, and the solar plexus.

We intensify the consciously perceptible physical exchange to "our intellectual level" by pressing on the root of the nose with the finger. The pressure should initially be very strong and then lighter as it fades. The skin can become red. The perception normally can also be felt intensively minutes later.

The Position of the Trigger Points and Zones

Trigger Points and Trigger Zones

If you have ever experienced the relieving and relaxing effect of an acupuncture treatment, you will know what I am talking about. It has been scientifically proved that relaxation is achieved by the body releasing *endorphins*, which cause this effect, at these specific places. In this process, endorphin is the antagonist of the tension-promoting adrenaline.

Through the treatment, the zones surrounding the tension are from freed of blocks and well-being can be perceived within the *entire* body. In his autogenous training, J.H. Schultz also described that even the relaxation of a partial zone like the "heavy arm" or "the warm hand" attains such an extensive effect that it also spreads throughout the entire body.

In the illustrations on the following pages we see the distribution of the trigger zones and points which we treat; they are also accessible for the anatomical layperson. These are key positions in our body which trigger relaxation when they are simply touched.

What exactly is a *"trigger point"* and a *"trigger zone"*? A trigger is a mechanism which—once activated—creates an effect? When I press the trigger of an alarm, the siren goes off. In the medical field, this expression indicates a point (or a zone) which actuates a process at a place which is further removed. When I inject an analgesic at a certain point, this leads to a relief of pain at the correspondingly distant or inaccessible point.

When we touch one of the points shown here, the hormone to which there is a connection is released. When I put my hand on the skin above the ovaries and sensitize them through con-

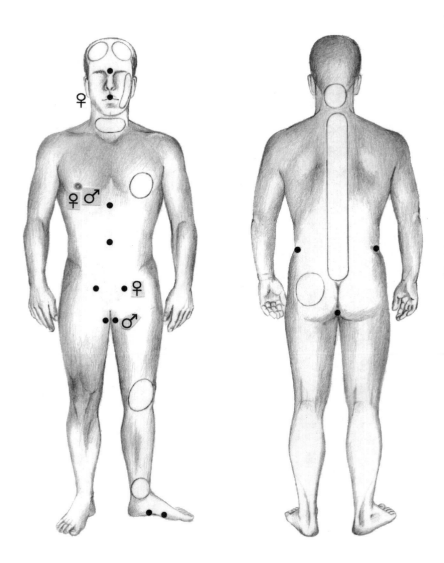

The human body with the most important trigger points and zones, in as far as they are of significance for endogenous induction

centration, the "wonder" occurs. Try this yourself. You will sense and feel how the ovaries are becoming "warmer."

The warmth of the hand comes from an endogenous warmth which is developed from within. At this moment, the estrogen is already released and all the corresponding effects manifest themselves. In the process, certain hormones themselves become trigger hormones. This means that the vagina becomes moist, the lips of the vulva swell, and the cheeks sometimes turn red: this all happens without a detour through an erotic situation.

When I put my hand on the stomach area, where the solar plexus is located, I have an effect on it and can feel it. It becomes noticeable, first becoming slightly warm and then more intensely warm. The person being treated has the sensation of feeling "a warm egg" there. The solar plexus is the most significant trigger point leading to the release of endorphins. We initially also feel a pleasant warmth in the surrounding area, as if we had drank a warm cup of tea. And now we can put the finger directly on the right point and give our trance journey free reign in the world of memories or visions of the future. It was namely the solar plexus that "ached" so much when we suddenly saw the love of our life. The treatment can be accompanied by the corresponding images. The solar plexus aches in all intensive emotions, in panic-stricken fear as well as stage fright or the excitement of being in love. The sensitive person perceives his or her feelings at this point.

We also find trigger points along our spinal column. The treatment of the nape area evokes particularly pleasant feelings and relaxation. When our cheeks are stroked, something different happens than when we try to create sensations in the body through the soles of our feet.

Neither extrasensory powers nor special knowledge and experience is necessary; we must simply just become sensitive—even if we need some exercise in order to do so. It is always important that we hold our hand in such a way as if we wanted to stroke a child and that we sensitize our skin so that it becomes a pure contact organ. Our skin is namely our largest contact organ

The Skin as a Contact Organ

Two of the many functions of our skin are particularly interesting. It is the organ through which we define both our boundaries and our contact. The self stops with the skin and the world surrounding us, the "you" and the others, begins. However, with about two square meters of surface, it is our most important contact organ. Every stimulation at any place is directed inwardly to a certain related organ. This occurs through a direct change of the electrical skin resistance, which can be measured in experiments today, but was already tested by C. G. Jung with his "association experiments."

The verbal expression of "I love you" causes an emotional perception which is unbelievably weaker than what we can express with caressing, stroking, or an embrace, which actually has the same meaning as these three words.

If the skin is the external expression of deeper emotional levels, we can find the way to deeper zones which are difficult to access.

An apparent example of this is blushing. Someone turns red because thought processes of which he is ashamed occur within his subconscious. The skin gives us away; it always tells the truth about what is happening within our psyche. This also become apparent during puberty because the skin reflects the changes in the psychological state.

In this sense, the entire skin is nothing more than one trigger zone. However, the *way* in which it is touched, whether it is an embrace or the touching of a certain painful place, is also important.

and, as such, it is one large trigger zone. When it is properly stroked—no matter where—we will feel and see the reaction (see box on page 48)! We recognize the enormous amount of possibilities for this because of blind people who can practically "see" with the skin and the help of their tactile cells.

Introduction to Endogenous Induction

Restoration of Hormonal Harmony

Endogenous induction includes a series of exercise sequences in order to 1. switch from tension to relaxation, 2. balance the contact between mind and body, and 3. reestablish hormonal harmony.

A large portion of the psychosomatically caused disturbances which limit our feeling of well-being can be traced to the over-proportional and one-sided use of our intellect. It is usually easier for us to solve difficult logistical problems than properly deal with our feelings, to sense and directly express them.

In many cases, the feelings are surrounded by taboo zones and filled with fear. This is why most of the newer therapies do not address the intellect, but instead try to find access to a person's understanding and change through the emotional life. Today, psychological problems, particularly neurotic disturbances, are much more likely to have their cause in the mind than in the "heart," which is permitted too little space.

One task taken on by endogenous induction is the restoration of hormonal harmony. Almost 2,000 years ago, Claudio Galenus already saw the cause of illness in the disturbance of hormonal balance, which is quite customary in our age. The principles of this great physician are still valid today in part and had defined medical science up until the late 18th century. A large portion of the various forms of neurosis has its cause in an overstraining of our "highest function," thought and the intellect. Thought is naturally the primary work of our brain. When thinking, we use up vast quantities of energy, no matter whether we are dealing with thought processes which are conscious or those which are not.

There are many people who believe they are not thinking at all at certain moments. At the same time, their brain is working unconsciously and accordingly uses up much energy. This particularly happens when the conscious grappling with problems is avoided and everything that is felt to be negative is suppressed into the subconscious—until it is like a barrel filled to the brim which can hold no more and, with the smallest additional drop, threatens to overflow or even does so. All of the problems that we evaluate as negative and try to get rid of through suppression usually "get us down " much more when they come from the subconscious. That we are tired in the evening without actually having done anything special is a very typical example of this subconscious mental exertion, which uses up much of the available energy.

The cause of tension has a physiological basis. In order for us to be able to think optimally, our brain must have a good blood supply for the necessary amount of energy to be provided. We achieve this increased blood supply to the "head" by tensing our muscles. This tension of the muscles in the body reduces the volume of the blood vessels and increases the blood pressure. The vasoconstrictive adrenaline has the same effect.

As a comparison, let us take a look at what happens during something like sexual excitation: there is an increased flow of blood to the genitals and we can therefore no longer think as well.

Each of us occasionally has the desire to simply turn off the process of thinking, particularly when the thoughts are unpleasant. When we have performed much mental work during the entire day, it would be absolutely relieving, as well as healthy, if in the evening we would let go of our thoughts for once. The blood could then freely flow to the other organs (heart, liver, stomach, kidneys and so forth), which all have much to do in order to process the toxic substances and waste products that have accumulated during the day.

It is remarkable how few people are truly capable of consciously turning off the wheel of thoughts for once in order to, for example, enjoy their meal in peace and quiet, have an undisturbed exchange of feelings—to be a human being and not an intellect. Which of us can simply sit in the outdoors, enjoy the sunset and listen to the wind, without being distracted by the disruptive rehashing of thoughts? Which of us must reach for a (harmful!) glass of alcohol or a convenient and supposedly helpful psycho-drug in order to calm down?

The too extensive and often all too worthless thought activity also follows us at night, when we would actually like to sleep undisturbed in order to recuperate. This tension also shows up in dream life in the same manner. The healthy dream has the task of balancing out the tension fields built up in the brain during the day. We only remember the content of a dream when its intensity extends beyond the threshold of our memory. Much too often we feel ourselves to "just" have had too little sleep, but we are mentally more despondent than before we went to sleep; in this case, although the body has organically regenerated itself, the brain has not. The "thinker" needs more than the 6 to 7 hours necessary for the organism; the brain should rest for 8 to 9 hours in order to remain efficient!

Someone who is "nervous" is a person with an overly active thought life. We can imagine what would happen to our heart if we would overstrain it in the same manner as the brain. The nervous person no longer takes the time to work through the problems that occur one after the other. Little problems then turn into problem areas which no longer can be worked through at all since the access to them is blocked. An energetic congestion occurs.

There are various ways of getting out of this situation. We can reduce our sensibility with respect to the perception of problems to the extent that we become tough and insensitive or we simply suppress problems out of the memory bank, to which we can consciously fall back upon. But this can only occur as long as there is still "room" there. The younger a person is, the more simply this occurs. But with increasing age, the amount of data gradually begins to overflow. With time, the memory can become so overloaded that it no longer can completely fulfill its function. Then thoughts emerge—often against our will—from the memory bank, the ability to concentrate becomes reduced, we can no longer properly enjoy life and have finally even forgotten what joy is. Life becomes senseless. Psyche and body no longer play along and become "sick."

There is an old saying which goes something like this: Only the simple-minded person, who doesn't think much or is not capable of making problems for himself, is happy. Time and again we meet people who can be envied for the grace of this happiness. They are less susceptible to "nervous" types of illness because they have few problems and can use their free capacity for mastering things as they come up concretely in a simple and direct way.

When the "nervous type" becomes a victim of his hyperactive thoughts, this either occurs consciously because it is a part of his socialization to scrutinize and exercise control over everything or unconsciously because the thoughts simply stream into him without him being able to stop them.

The equilibrium between mind and body becomes unbalanced. The more a person is in a state of balanced mind-body equilibrium, the less neurotic he or she will be.

This is why the first task of endogenous induction is to achieve a pleasant state of relaxation. The second task is to restore the body-mind contact and thereby switch fro thinking to feeling. The third task of endogenous induction, the hormonal equilibrium, then occurs completely on its own.

We know that purely physical activity, such as jogging—which means running to the limits of one's own capability—limits the capacity to think. The blood is drawn from the head into the muscles and the flow of thoughts becomes calmer. The rhythm of running, like all rhythmic movements, releases endorphins. This is also the reason why a person can become addicted to running. The body's own morphine-like endorphin is accompanied by a feeling of exaltation, to which one can become addicted. If a person is prevented from jogging for a certain period of time, withdrawal symptoms may even occur.

Endogenous induction acts in a different manner. It achieves a stronger physical feeling and a switch from thinking to feeling through activating various body zones. By producing strong perceptions which come from the body, it interrupts the undesired train of thought for a period of time.

Something similar to "tactile hypnosis" occurs in endogenous induction: by satisfying ourselves with physical perceptions, almost our entire thought activity comes to rest.

When we touch a special point of the body in the proper way, we thereby create a correspondingly strong perception. And—this is the next step—if we can have an effect on a gland through this point or this zone, we can achieve the release of one or more corresponding hormones. We have mentioned the example of "hand contact above the ovaries": the skin temperature increases in accordance with the greater supply of blood and of "energy" in general. The perception is concentrated upon this area: estrogen is released. I am certain that this change can also be measured. In any case, the corresponding effects such as redness of the cheeks, a need for contact, and extroversion appear very quickly. I have often experienced that finger pressure on the solar plexus is felt to be a soothing "penetration of a warm object," which releases tension. Pressure on the base of the spine is just as effective: "As if a warm ray would stream upwards along the spine." The tension is always relaxed when this is done. At best, we distinctly feel the entire spine. Afterwards, a pleasant side-effect can be seen in an erect posture.

The Three Different Types of Treatment

The quality of our relationships is very important for our personal well-being. At the same time, we differentiate three different levels of relationships:

1. The Relationship To Our Self

At this level, we differentiate three layers within our own personality: thoughts, feelings, and instincts and their harmonious expression. Each of these levels should be freely experienced and lived out. As long as there are no conflicts between thinking, feeling, and acting, a person lives from his or her center. The exercises for individuals are dedicated to this aspect.

2. The Interpersonal Relationships

In the broadest sense, this is the *intimate sphere* which can develop between ourselves and another person. This can be a lover, mother, father, or also a friend with whom we have a platonic relationship.

3. The Relationships With Our Social Environment

These are the relationships to people with whom there is no intimate contact. These relationships in the occupational and social area are contacts which can also influence our well-being.

Freeing the Psyche of Inhibitions and Blocks

Exercises Which Can Be Done Alone

When we feel free to follow more of a feeling of desire than duty and grant ourselves an appropriate way of expressing ourselves, this is a sign that there are no disturbing inhibitions within us. Inhibitions occur when there is a lasting conflict between the authority which controls us (a moral-ethic, trained or acquired structure) and our natural powers and/or instincts. The free expression of our feelings, which means the intimate design of our life, is also important for our authentic feeling for life.

Our exercises are aimed towards achieving a relationship to our body which is free of prejudices. A part of this is the inclusion of all body parts. This is easy to say—but there are only very few people who really have an open relationship to every part of their body to the same degree.

The path to completely accepting our body goes through the heart. When we listen to our heart as the symbolic center of our feelings, the inhibitions in the expression of feelings are relaxed. Then we are willing to also recognize the sexual zones as organs of equal value and allow the full perception of all sensations in this area, thereby freeing our instincts of the taboos which still exist. Here "instinct" means permitting something completely normal to take place—just as we allow ourselves to accept the sensations from our stomach which signal hunger—a completely normal need that wants to be satisfied because it is essential for life. Not all physical needs are essential for life, but they can considerably increase the quality of life.

The following individual exercises are coordinated with each other in such a way that we can once again understand—even better, feel—the most frequently forgotten body zones in their totality and accept the body in its entirety. The exercises should help let go of any extremes of an intellectual, emotionally-determined or one-sided instinctually-determined overemphasis and thereby help to bring the harmonious interplay of the hormones back into a state of complete equilibrium.

Deepening the Contact Between the "I" and the "You"

Exercises for Partners and Friends

The second exercise series is designed so that they can be done together by two people. We can work through this exercise sequence with our partner or an intimate friend—precisely the people who are close to us.

Once the first step to a good body-mind contact has been taken, we will already have a much more intensive inner harmony. This alone often causes our interpersonal relationships to change. We can arrange our partnerships far more intensively in a way which is pleasant for living together. Our feeling for life, and thereby our happiness, is intensified in that thinking, feeling, and acting are in a more harmonious accord with each other.

Words that express our love and our understanding for each other should lead us to a deeper dialog, a true contact on the mental level. This contact gives us the greatest fulfillment when it is so intensive that it actually touches us.

There is a big difference between dialog and a "deeper dialog." In the first case, we are speaking only about facts such as the weather, for example; in the second case, a partner can reveal his or her feelings with or in addition to the facts. For instance: *"When we met in Venice two years ago, you had on a red dress, I saw your shape, but above all: your face radiated. My heart beat more quickly, I was moved and hid the flowers I had for you behind my back like a bashful little boy...I will never forget this moment."*

Even in this image, in the recounting of this encounter, an increased release of hormones occurs. In the process, something is certainly moved in the area of the heart and perhaps something is also moved—still quite impassively—in the erotic zones.

Now we have taken the second step and start to express our feelings by telling about ourselves on all levels and particularly expressing what is close to our heart.

Although this already occurs when expressed by speech, we can intensify it through skin contact which is emotional and tender . The entire body will also resonate at the same time.

Following the syntonic state, a harmonious matching of impulses, on the mental level, we come to the syntonic state of feelings and then ultimately to that of the entire body.

Only when mind, soul, and body become one can there also be a complete exchange on all levels during sexual experiences. Each of us *alone* is not in complete balance and needs the other person in order to become complete and whole.

We can start the exercises by each doing the individual exercises together and telling each other about how we experience them.

Then we can start with the partner exercises, through which a more intensive syntonic state of the bodies occurs. This syntonic state is the basis for a profound sexual exchange—which we will not delve into in this book. Without this exchange, sex remains gymnastics in which testosterone and/or estrogen is reduced and desire is satisfied only on the physical level.

Only when we live out a relationship with the partner which experiences the mental and emotional elements are hormones built up before and during the sexual exchange. And only as long as hormones are built up does the relationship remain a lasting one and withstands the desensitizing of feelings for each other through the force of habit. After such a complete sexual exchange, a person feels "satisfied" and is still filled with love for the partner.

This naturally does not exclude sexual intercourse which is only physically demanding but still exciting at any rate, particularly during the youthful excess of vital hormones. But if this becomes the rule, it soon loses its appeal and no "technical" intensification can remedy the rising boredom.

Feeling the Contact to the Human Environment

Exercises Which Can Be Done in a Group

What is meant here is what Erich Fromm already described with "...whoever loves a human being loves the entire world." As social creatures, human beings need social contacts and must be able to perceive how the people in their environment turn towards them.

The group exercises convey a pleasant feeling of absolute security. Here we also experience ourselves purely physically as the center of the world, which we—seen from the philosophical perspective—actually are. Through this strong experience, the "I think therefore I am" becomes a "I feel therefore *we* are."

Taking is an important component of group exercises, whereby we also give something to everyone else.

There are very few families today that still have a deeper contact which permits love, affection and mutual interests to be exchanged intensively.

It is interesting that it is often easier to do these group exercises with *strangers* whom one has not seen beforehand and also will usually not be seen afterwards.

The feeling of being the total center is conveyed when we lay on many hands. Hands, which we feel on the greatly varying zones of our body, relax both the psyche and the body to the same degree and impart to us the feeling of being part of a world full of love. It is an uplifting and enduring experience.

In endogenous induction, there is much more that happens than just the laying on of hands. It is a perception and exchange of energies. This form of treatment requires our loving attention and sometimes also patience when we treat people who have very much lost their mind-body equilibrium, as well as by people who are strongly intellect-oriented and nervous.

At the same time, a skilled trainer can already let his or her clients feel an extensive effect, which they alone would need somewhat more practice to achieve the very first time.

We will now begin with the various exercise programs here. The choice of exercises is extensive. You should try all of them at least one time in order to ultimately discover for yourself which ones are the most effective. Afterwards, you can use the time you set aside for yourself for the exercise positions which evoke the greatest degree of well-being within you.

We will change with the exercises, growing in our personality, attaining a harmonious self-expression and having more personal magnetism. We can quickly perceive the positive changes within us—even when the process which has been triggered tends much more to be one of slow and constant growth than one of sudden discovery.

The Exercises for Endogenous Induction

Switching from Tension to Relaxation

Reducing Adrenaline and Building Up Endorphin

If we once again imagine a tense face and a cheerful, relaxed face in front of our inner eye, we have a rough idea of the great change which can occur within us when switching from tension to relaxation.

Although the relaxation exercises bring a great deal of relief, it is not possible for many people to immediately start with the actual exercises of endogenous induction from a relaxed situation.

Before describing the individual exercises, we have compiled the exercise sequence in a summary. This makes the course of the treatments more clear, and it will be easier for us to remember them later, even without the guidance of the book.

Treatment Summary

1

Relaxing
movement exercises:
running in place;
dancing; rhythmic
movements

2

Contact with the earth

3

Moving according to the needs of the body

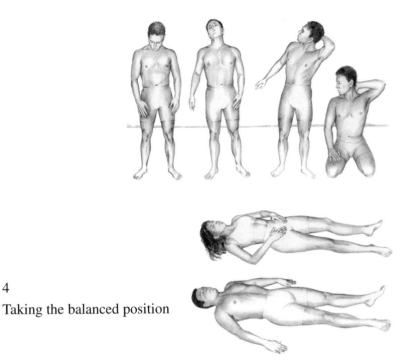

4
Taking the balanced position

5
Breathing exercises

6
Looking inwards and exercising the powers of imagination by visualizing the individual parts of the body

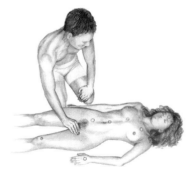

7
Heightening the skin's
sense of touch

The Relaxation Exercises

1
Relaxation Movement Exercises

Running in place, dancing, rhythmic movements

Relaxing muscle tension

Note: In order to achieve a pleasant relaxation of the muscles—the precondition for endogenous induction—there are various movement exercises available which are easy for you to do at home.

Instead of jogging outside, you can simply run in place or dance to rhythmic music that appeals to you.

Simple gymnastic exercises are also suitable for relaxing the body. You can either freely improvise or use simple sport equipment, such as that offered everywhere today. The most important thing is that it's fun.

The sole decisive factor is the intensity of the movement exercises: in the process, the pulse should be accelerated by about 10%. This already happens when you start to sweat during the movement.

In the rhythmic movements, endorphins are released as well. The effort is then directly perceived with a heightened sense of well-being.

2
Contact With the Earth

The feet—barefoot, if possible—feel the contact to the ground; we can stand in the balanced position* or also stretch the hands up to the sky or clasp them at the nape of the neck.

Creating contact with the earth

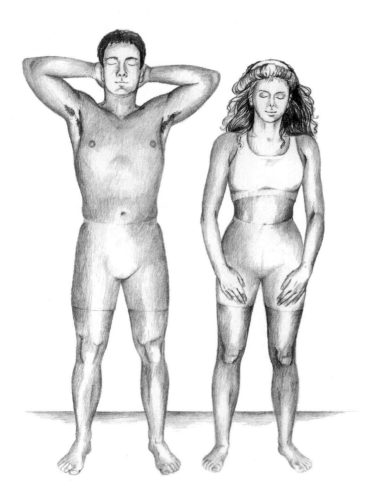

Note: Today it is not even possible for people everywhere to stand on the "earth" with bare feet. For this reason, you sometimes will have to assist with your imagination here—with the vision of a dewy meadow, of warm sand on the beach, or whatever else appeals to you.

The imagination helps us even further: You will become one with the earth, flow into it, and let all the negative energy flow away on its own and into the ground.

But you will also feel the support under your feet, upon which our core self can find its equilibrium.

In addition, you can now let your body become "heavy." This perception occurs almost on its own. If you concentrate on it, you can feel how the blood moves from the head in the direction of the feet.

As often as you have the opportunity, and naturally in particular when you actually are outside in nature and standing on real earth, you should grant yourself this experience: simply take off your shoes and enjoy the contact with the earth.

Some of you may perhaps remember a happy childhood and times when you still went barefoot.

* The balanced position is described on pages 72 to 73

3
Moving According to the Needs of the Body

Moving the body to its own melody

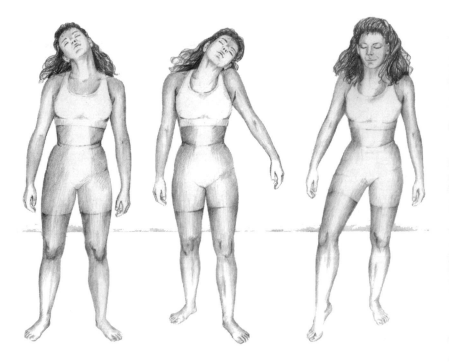

Note: This exercise is the easiest, even if it appears to be more difficult at the start than all the others because we have forgotten how to follow our body.

But if we listen to our body we will realize what it needs!

To come into harmony with the body is most difficult for those who find themselves in disharmony with their body and can perceive

no melody, no body rhythm of their own, because of a chaotic inner life.

But even this is not a problem if you calmly stand still! In such cases, it suffices to simply wait, listen to the first clear signal and perhaps help out a bit using the will at the beginning: with a circular motion of the hand or a slight, circling swaying.

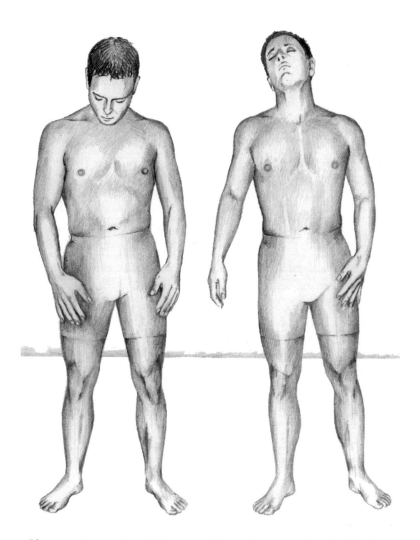

If we put ourselves into a position of rest and patiently listen within ourselves, it is possible for us to feel the movements required by the body; at first, it can be a "pull in the neck" according to which we turn the head, then the shoulder wants to move upwards, the knee forwards, or we feel the need to kneel down.

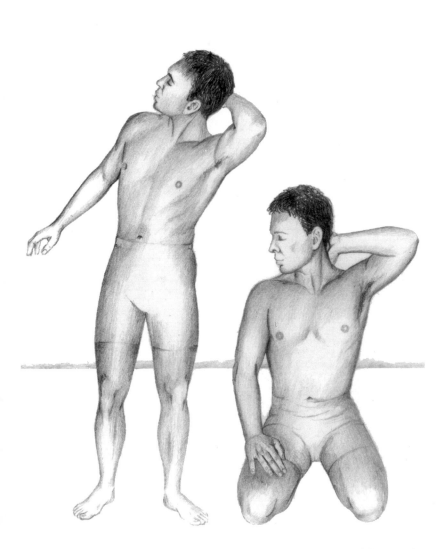

Things can also start to rage within us; then we hit a pillow and work off the tension as much as possible. At the end of this exercise, we should always take a pause, particularly when we have released many emotions and want to restore our accustomed equilibrium.

4
The Balanced Position

Lie down and assume the balanced position, looking inwards and without a goal. Passively perceive the impressions that arise within you.

Listening within

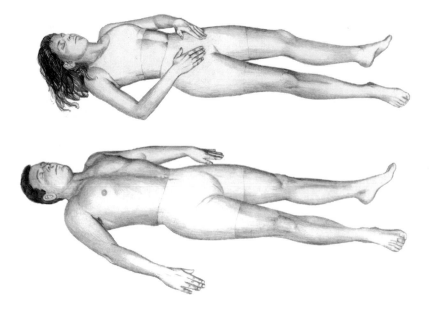

Note: Here you see the balanced position. The body lies on the back in complete symmetry, and the stretch muscles and pull muscles are in a balanced position. When there is no more tension in the body, the hand should be *flat* and completely relaxed. It is important to be sure that the hand does not lie in a position which could be expressive of something like drawing closer or holding tight.

The tips of the toes are pointed slightly outwards. We try to achieve a perfect symmetry in this position and observe our body as if we could view it from a point which lies above and below us.

The concentration on our body calms down our thoughts. You should completely forget to think and just see clouds passing above you and feel a light breeze. This is the ideal body position—even if it is not for everyone at the start since chronic tensions which have already slightly deformed the body can become painfully noticeable in this position.

You may possibly feel the urge to cross your feet. This feeling only lasts until the relaxation of the body has also reached the psyche.

Although the position may then be a bit troublesome at first, you will perfectly master this initial position sooner or later. The momentary inhibitions and tensions show up during this "meditation."

Incidentally: a curled-up sleeping position is due to tension of the stomach muscles. If you succeed in falling asleep in the balanced position, the ensuing sleep will be deeper and more relaxing.

In the balanced position, the mind and body confront each other. There is an intensive body-psyche encounter—if you open up here, you will encounter your own truth.

5
Breathing Exercises

Observe the individual breaths in the breathing exercises, then inhale deep and long, saturating the body with oxygen until a slight feeling of intoxication occurs.

Feel the basic rhythm of life:
Breathing in and out

Note: "There are two kinds of blessings in breathing ..." As a matter of fact, breathing is certainly the most important organic function through which we can achieve relaxation.

The vital oxygen supply for the entire body, particularly the brain, is dependent upon it. Without a sufficient supply of oxygen, we cannot think. If we interrupt the breathing rhythm, the first brain cells will die shortly thereafter and the entire organism dies very soon afterwards. Despite this fact, the great majority of human beings do not breathe in a manner optimal for the supply of oxygen. Even the difference between chest and stomach breathing is not familiar to everyone.

Influence on the body: Seen in a physiological perspective, it would be good to use the maximum volume of our breathing potential. When we observe our breathing, we feel how the stomach and chest fill up with breathing air. Chest breathing is normally neglected. Breathing regulates itself. If our body needs oxygen, we involuntarily breathe more strongly and intensively.

When we follow our breathing, we feel how the air is cooler as it flows into us and warmer as it flows out. In the process, we become aware of a perceptible physical relief.

Influence on the psyche: There is a deep relationship between tension and relaxation and breathing in and out. If we are tense psychically, we breathe poorly and fear can take away our breath. We "simply cannot breathe" in a room which we find unpleasant.

While breathing, we can imagine how the closer and more distant surrounding world flows into us. We actually change the entire world with every breath. Breathing is the bridge between the self and the environment—we cannot isolate ourselves. Here we clearly see the necessity of contact, but also our dependence. Breathing in and out is the basic rhythm of our life: inhaling brings us tension and breathing out perceptibly relaxes; inhaling is taking and breathing out is giving, inhaling means making contact and breathing out means letting go, inhaling is limitation and breathing out is release. After the work has been done, we heave a sigh of relief and thereby relax.

Every person begins life with the first breath and becomes independent because of it, free. We also end our lives with the last breath and free ourselves from the limitations of this life.

Every room has a window. We can open it and heighten our entire mental-psychic activity with the fresh air. The hunger for air is simultaneously a hunger for freedom and free space. We could also say that the breath is not within us, but rather we are in the breath, as if in an endlessly large womb.

Abandoning oneself to these considerations means bringing the self back into its real dimension. We once again become normal people and are relaxed, perhaps even happy as well.

6
Looking Inwardly and Exercising the Power of Imagination by Visualizing the Individual Parts of the Body

We once again bring ourselves into the balanced position, being sure that the body comes into a harmonious equilibrium and that we are not disturbed by any undesired perceptions.

Then we begin to visualize the various parts of the body in detail, one after the other: knee, nape of the neck, stomach, hand, forehead, as well as the heart, lungs, abdomen and the sexual organs.

We train ourselves in sensitive anatomy, becoming better acquainted with the body and achieving the desired state of partial hypnosis through this intensive contact. We completely concentrate on visualizing our body in detail. Where stronger perceptions take form, we focus upon them with our attention. As a result, the respective organ receives a greater supply of blood.

Feeling and seeing the individual parts of the body

Note: With the preceding exercises we have achieved a partial relaxation and have also found a better contact with our body at the same time: the endogenous perceptions, which originate in our body, overshadow the exogenous, those which come from the outside, such as noises and all other sensory impressions as well.

It is the objective of the exercises that we feel ourselves to be a unity and can completely concentrate ourselves upon the perceptions directed inwardly. If we now turn to the knee, we can practically see it in front of our inner eye, we feel its bones and cartilage. (If we are not very familiar with anatomy, an anatomy atlas can be very useful for helping us understand the body.)

During this, we should let go of all the prejudices which we have against certain "hidden" parts of the body, the so-called taboo zones. The relationship to the penis and vagina should be no different and just as natural for us as the relationship to the nose and mouth or at least equal to it.

It is precisely these concealed zones which are psychosomatically significant areas where much potential to work through can be dammed up. Perhaps we will even be astonished—and there is certainly much which is astounding.

7
Increasing the Skin's Sense of Touch

The support of a second person may be necessary for this exercise. Furthermore, it's also a great deal of fun when two people do it together. We assume the balanced position, our partner sits at our side and very gently places coins of various weights on the skin of the different parts of the body. It is our task—with our eyes closed—to state the exact position of the coins and their weight (for example, penny, dime, quarter), how much warmth they create, and so on. There are infinitely many ways of playing this and numerous possibilities of increasing the perception. We only have to think of blind people who are capable of "seeing" not only with the hand, but also with the skin

Increasing the sensitivity of the skin

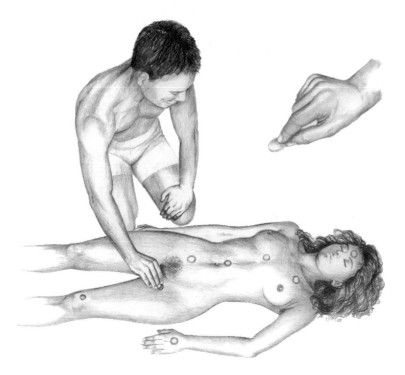

Note: On the one hand, we increase the sensitivity of the skin in this exercise; on the other hand, it makes it possible for us to discover blocked body zones.

On some places of the body, usually the erogenous zones, it is completely possibly for psychosomatic barriers to exist. A young woman whose breast was touched commented: "It was as if the breast of another person next to me was touched; I felt like a piece of wood at this spot."

Every body zone can fundamentally be "turned off." Unpleasant memories, fears of illness, and many other things can be stored there. Seen energetically, the body is a mirror of our perceptions and a depot for our memories. Here we can expose the causes of psychic inhibitions so that they are once again accessible to conscious treatment.

Releasing Inhibitions and Blocks

Exercises Which Can Be Done Alone

How to free the psyche of inhibitions and blocks, intensify the contact between mind and body and restore the hormonal equilibrium.

Survey of Individual Treatment

A) Lying on the Back

A1

Left hand to right foot
(and the other way around),
massage sole of foot

A2

Left foot to right knee
(and the other way
around), fold hands at nape
of neck

A3

Knee to chest—grasp the
ankles with the hands
and press

A4

Stroke the inner side of the thigh from the knee to the groin with the fingertips

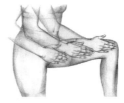

A5

With light pressure, press the hands on the groin area, the man touching the testicles and the woman touching the vagina

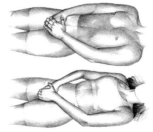

A6

Exercise for the man: Press on both sides of the groin with the fingers

A7

Exercise for the woman: Press above the pubic bone with the hands on top of each other. Afterwards, let the hands glide upwards with decreased pressure.

A8

Press lightly on the navel with the fingertips

A9

Place the hands on the
solar plexus

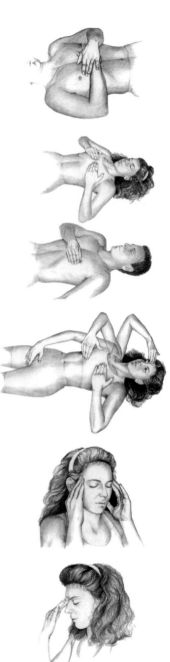

A10

Massage the chest,
breast,and nipples (exer-
cise for man and woman)

A11

Stroke from the root
chakra to the forehead
chakra with the fingers

A12

Stroking the facial muscles

A13

Strongly press the root of
the nose with the index
finger and middle finger

A14

Put one hand under the head
and the other on the forehead

A15

Touch the temples with the
fingers, hold the ears closed
with the thumbs

A16

Massage the skin of the head

A17

First put the hand on the left
hemisphere (side of the head)
and then the right

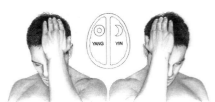

B) Lying on the Side

B1

Press together the entire spinal column through the coccyx and the first cervical vertebra

B2

With one hand at the nape of the neck and the other below the knees, press the forehead against the knees

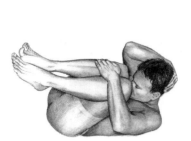

C) Lying on the Stomach

C1

Grasp the ankles and press the heels against the buttocks

C2

Massage the buttocks

C3

Place the hands around the waist and stroke the kidneys

C4

Place the head on the elbows and press against a pillow

D) Erotic Movements

D1

Consciously imitate the movements of sexual intercourse

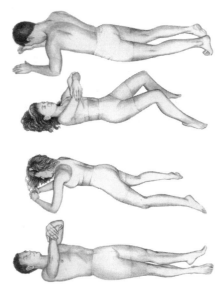

Individual Exercises

A) Lying on the Back:

A1
Left Hand to Right Foot; Foot-Sole Massage

Foot: **Effecting the organs through the sole of the foot**
Ankle: **Feeling secure**

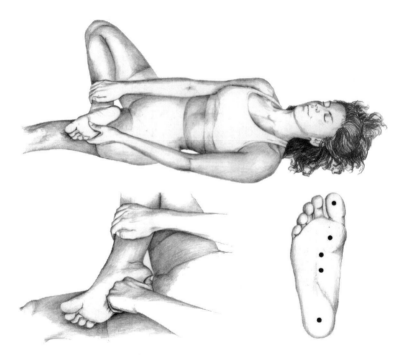

Note: If you have ever had the soles of the feet massaged, you will be familiar with the sensations which are evoked in completely different parts of the body. The massage is particularly pleasant when the person performing it is familiar with the individual reflex points.

We relax during the massage and observe how a light flow of energy from the foot to the organs becomes perceptible.

Our illustration shows the points on which we should exert a stronger pressure.

A2
Left Foot to Right Knee and The Other Way Around—
Hands Folded at the Nape of the Neck

Knee: **Freely moving forwards**

Nape of the neck: **Creating connecting thoughts**

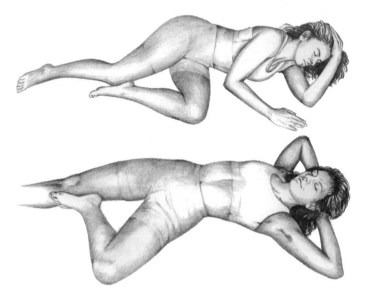

Note: In this exercise, we connect two important areas of the body, the knee and the head. The knees, which has a psychosomatically interpreted significance of one's own progress, the steps we take into our own future, is connected with the head, which dominates our conscious space to move.

We bend a leg and put the sole of the foot on the other knee at the same time. We place the hands at the nape of the neck and thereby include our head in the literal sense.

Pain in the knee is often the result of the inability to further develop ourselves, not desiring or not being able to do so, and very frequently has a psychological reason.

A3
Knee to the Chest—
Grasping the Ankles with the Hand and Pressing

Enjoying the entire body and yourself

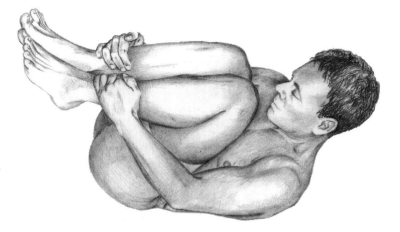

Note: We see and feel our body from the outside—with a pleasant distance. We embrace and stroke it while curling up. Curled up in this manner, we feel how the large zones of our body enter into skin contact with each other. In this position, the forehead touches the knee, we feel much more at one with ourselves than we usually do. This exercises is different from the previous ones because we create a natural tension.

In doing so, our own muscle power becomes perceptible.

A4
Stroking the Inner Side of the Thigh from the Knee to the Groin with the Fingertips

The intensive perception

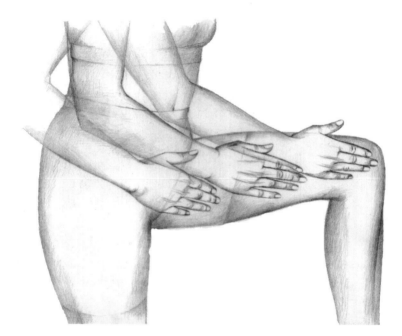

Note: We normally seldom touch the inner surface of the thigh. During this exercise, the body should remain still (as in the balanced position) and only the hands should move. Perhaps we will have completely new sensations while doing this. At the beginning, it can be a type of tickle—a very strong transference through tactile cells which are seldom touched—which then turns into a gentle excitation. The inner side of the thigh becomes warm. Precisely this zone is very sensitive since it is the arousing pathway to the genital area.

The inner side of the thigh is a strongly taboo zone, particularly for young women. Even the hand of a person to whom they are very close initially creates an unpleasant excitation which is often felt to be dangerous. When we run along the inner side with the finger, we easily discover the right line and the points along it which are particularly stimulating, where we then pause for a while.

For a man, this exercise can lead to a nervous discharge of the testicles. The woman very strongly feels the psychological effect in particular.

A5
The Hands on the Groin Area with Light Pressure
(The man touches the testicles, the woman the vagina)

Reducing prejudices and building up a contact to the "dark" zones without prejudice

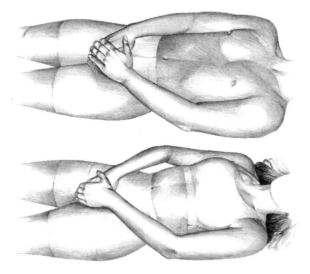

Note: Above all, you should have no false shame or fear of this exercise. We consider the sexual organs in the same way as any other organ. This is a challenge for many people, and it is important for them to accept it. Going beyond a taboo also means conquering your own fear.

The perceptions can be very strong. You can almost always feel a strong pulsating, and warmth becomes concentrated there. However, no erotic sensations should be evoked. The exercise attempts to more intensively open us to the perceptions from this area, to feel them less judgmentally, so that we can accept these organs as a living part of our body and develop a natural relationship to them. We perceive all the impressions, visualize vitality, security, and the pleasant feeling of the contact.

A6
Exercise for the Man:
Pressing on Both Sides of the Groin with the Fingers

Stimulation of the vital energy

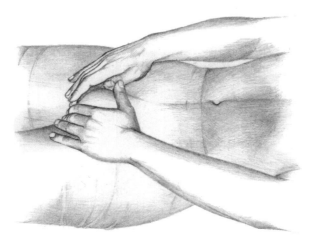

Note: The hands are laid onto the zones shown in the illustration. In the process, we put a more or less intense finger pressure on the inner side of the groin and do this completely according to our own feelings. Warmth is very quickly created when this is done. In doing this, we touch a point above the peripheral nerve that leads to the testicles.

This exercise increases the activity of the glands which release testosterone. By doing this, the ability to become aroused is strengthened not only in the erotic sense, but in general as well.

A7
Exercise for the Woman:
Press Above the Pubic Bone with the Hands Laid on Top of Each Other. Afterwards, Let the Hands Glide Upwards with Decreased Pressure

**Sensitizing the entire abdomen
and the ovaries through warmth**

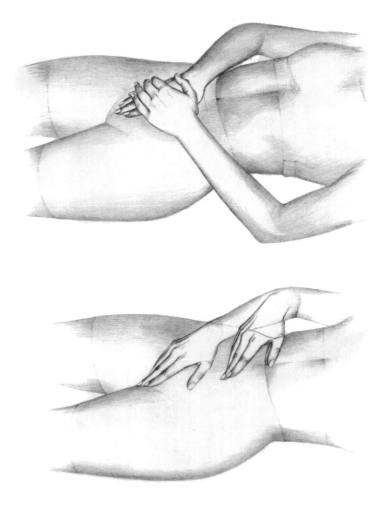

Note: We now make contact with the ovaries, which we perhaps have not perceived with such clarity, except during menstruation. They become warm through the laying on of the hands and can be perceived especially well for this reason.

An interesting process now begins. Secondary effects such as the reddening of the cheeks and increased moisture in the vagina occur. The physical changes are accompanied psychologically by a need for closeness—we become more extroverted and open.

We now move the hand upwards, stroking over the abdomen, to which we can never give enough caring attention so that it remains soft, flexible and sensitive.

It is very conspicuous that many women who draw back from sexual exchange soon gain cushions of fat on the abdomen, which are the equivalent of a protective shield.

With this exercise, we maintain our youthful liveliness and vitality. But it is also particularly helpful for women who have no sexual intercourse (or no longer have it). The positive effects of the estrogen released through the exercise become strongly perceptible here.

A8
Pressing on the Navel with the Fingertips

Returning to the source

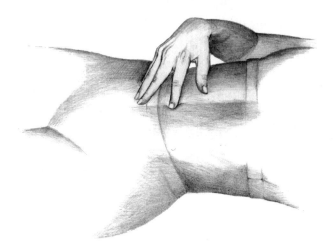

Note: We often forget that we also have a navel and that at one time we were connected with the source of our life through this navel. We lived in symbiosis with our mother, our paradise, from which we were expelled with birth. The umbilical cord was cut at that time. There can be moments when it is simply nice for us to return to this paradise in our thoughts...

The exercise securely leads us to a more intimate relationship to our mother and the Great Mother, our planet Earth.

A9
Laying the Hands on the Solar Plexus

**Feeling the sun within us and letting the feelings
of our first love come back to life**

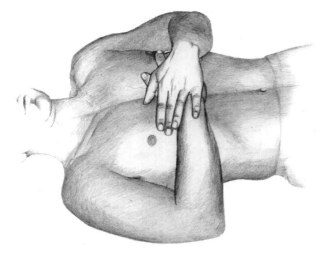

Note: The solar plexus is a network of nerves which is located
about a handbreadth above our navel. The main task of the solar
plexus is to regulate the digestion (stomach, appendix, large intes-
tine, liver ...).

We know how extensively these organs react to our psyche. The
relationship between the solar plexus and emotions is of particular
interest for us here. The name "solar plexus" is not coincidental
since the seat of our body's own sun actually is here. The solar
plexus radiates heat and indirectly releases endorphins since it creates relaxation as soon as we lay our open hand on this zone. We all
remember something like the time of our first love when the "stom-

ach" hurt. All strong feelings, whether they are joy, fear, dread or love, have an effect here.

Not the heart is the center of the emotions, but our sun. When we warm it by laying the hand on it, there is a pleasant effect on the entire body. Once we have felt the exact treatment point with the flat hand, we can quickly have very intensive perceptions through finger pressure.

A10
Massage of the Chest, Breast, and Nipples
(Exercise for man and woman)

A feeling of well-being flows from the chest

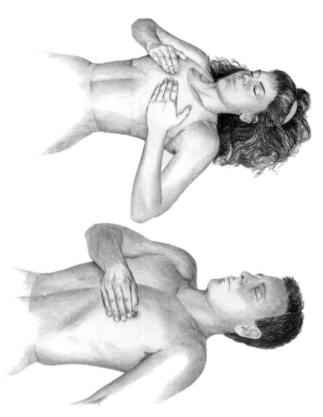

Note: We often forget that in the breast, mainly in the nipple, the man's sensitivity is the same as the woman's. If we hold the hand in the form of a shell over the breast, this releases a feeling of warmth which radiates out throughout the entire chest. Like the woman's nipple, the man's reacts to tender stroking. Treating it playfully creates strong perceptions, relaxes the chest musculature, and increases the sensibility.

A11
Stroking from the Root Chakra to the Forehead Chakra with the Fingers

Connecting feeling, thinking and acting with each other

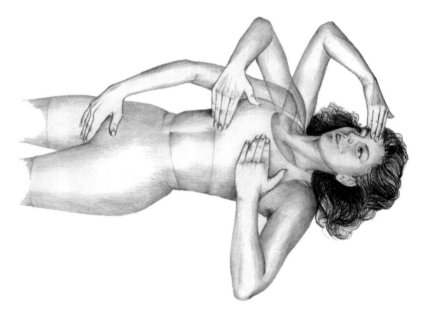

Note: We repeat this movement a number of times in alternating directions, from below to above and from above to below. We can briefly pause on certain points and thereby heighten the perceptions. This applies particularly to the beginning and end points of the movements. Through stroking motions, various body zones are connected with each other. The sensation created by the stroking is relaxing and can help us go to sleep. This can happen easily when we feel the effect with special intensity in the zone of the solar plexus.

A12
Stroking the Facial Muscles

Once again experiencing loving affection

Note: Our cheeks should always have been the center of attention for caresses. These zones invite us to stroke them. We can convey our love through them and take the first step to a greater degree of intimacy. Caresses on the cheeks evoke memories that are both strong and pleasant for almost every person. First it was the hand of the mother and of the father, and then the hand of another loving person, who stroked us.

Stroking with one's own hand brings back these moments, which have often been forgotten. Through fantasy and imagination, what we desire is created—the things which can be given to us by a relationship characterized by trust, responsibility, and affection.

A13
Strongly Pressing on the Forehead Above the Root of the Nose with the Index Finger and Middle Finger

Making contact with the mental level

Note: If there is a physically locatable access to our mental level, then it is the "forehead chakra". It is the spot where women in India wear a dot. This is the place we look at when we want to "get through" to another person.

The pressure should be particularly strong at the start so that the effect lasts for a while. We can even feel a brief pressure for some minutes afterwards. The spot can become slightly red and hurt a bit. During the exercise, we try not to think but instead visualize our hemispheres and perceive them as they switch from one half of the brain to the other. The effect of this often reaches as far as the cerebellum.

A14
Putting One Hand Under the Head and the Other on the Forehead

Holding the mind in the hands

Note: After the locally limited pressure of the previous exercise, we now *embrace* our mind very tenderly.

With this, we achieve a direct feeling of being protected or sheltered. After all, we are holding the part of the body in our hands which often appears to be the most important one to us. This embrace is very pleasant. We hold it for a while and enjoy the resulting perceptions.

This part of the body appears to us to be decisive for our life: The head is virtually the boss of our own body, yet we are *permitted* to take it into our hands and protect it for once, instead of being dominated by it.

A15
Touching the Temples with the Fingers and Holding the Ears Closed with the Thumbs

Perceiving the pulsation of the thoughts

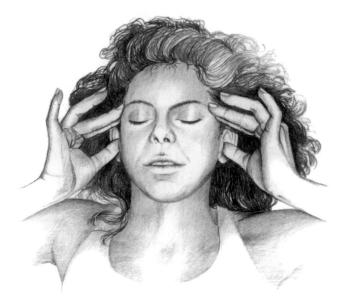

Note: We close the ear, and no more sounds coming from the outside can force their way into us—we can listen inwardly. Now the pulse becomes acoustically noticeable: Through the fingers which are gently placed on the temples we achieve a physical perception of the flowing, pulsating blood. With the help of this sensation, we can imagine the *work* of the brain and receive an idea of what is happening inside of it. Whether we are consciously thinking or not, our brain activity is unbelievably extensive—and we also want to feel exactly this fact.

A16
Massaging the Skin of the Head

Feeling the train of thought

Note: We massage the skin of the head with the fingertips and awaken within us the pleasant feeling that we sometimes have while washing our hair. However, it is more effective when we pull on the hair and thereby slightly "lift" the skin of the head. The result is an immediate improvement in blood circulation.

We literally "wake up" and experience how much the strength of the blood circulation is connected with the intensity and the alertness of our faculty of perception. Whenever tiredness overcomes us and the brain activity slows down, we can use this exercise. This also applies to situations like long drives and while we study or do other mental work.

First Putting the Hand on the Left Hemisphere (Side of the Head) and Then the Right

Activating the yin and yang energies

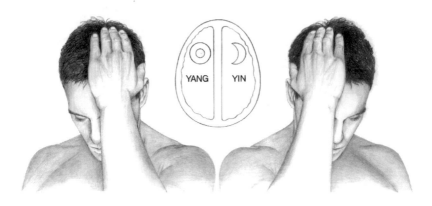

Note: The wisdom of nature is fascinating. The brain is divided into two "autonomous" halves, of which one controls the left side of the body and the other the right side. In this process, the peripheral nerves leading from the head to the body cross each other. This is why only one half of the body breaks down and the other half remains fit when there is a one-sided injury to the head. Every region of the brain fulfills special tasks. The left half is responsible for logical thought, the verbal power of expression, and the complex background of expression in the form of language. The right half is responsible for sight, fantasy, and so on, ranging into unbelievably specialized task areas. One example of this is that a modest group of a few million neurons (the entire brain has about twelve-thousand million neurons) has the task of processing visual impressions.

It is import for us that one half, specifically the right half, is responsible for the emotional world. Within it, we localize the yin values like "heart," moon, night, and subconscious. The left half of the brain has traditional symbols like sun, activity, masculinity, consciousness, fire, and so forth assigned to it.

We have discovered that the right side hurts when we are too much burdened by feelings, such as the pangs of love, worries, fears, and so forth. Headaches on the left side occur through intensively pursuing intellectual problems or tasks such as the solution of problems, studying, or similar situations.

This exercise familiarizes us with our brain functions: We first touch the left side and then the right side of our head and feel the warmth created in the process.

B) Exercises Lying on the Side

B1
"Pressing Together" the Entire Spinal Column Through the Coccyx and the First Cervical Vertebra

Perceiving the support of the self in the spinal column

Note: This complex exercise requires time in order to achieve the desired effect.

There are people who have no backbone—in the figurative sense, but their relationship to the spinal column is also poor in other ways. This exercise is particularly meant for these people. Before doing the exercise, it would naturally be good to have the muscles enclosing the spinal column massaged—as described in the partner exercise on page 132. The pressure on the coccyx can be very painful. But this pressure is generally felt to be like a warm wand moving upwards through the spinal column. This feeling of warmth can also be limited to one or two vertebrae.

The exercise is successful when we have the feeling of holding our spinal column in our hands. The positive side-effect of this exercise is a stance which is physically as well as mentally very upright.

B2
With One Hand at the Nape of the Neck and the Other Below the Knees; Press the Forehead Against the Knees

Collecting oneself and uniting the body

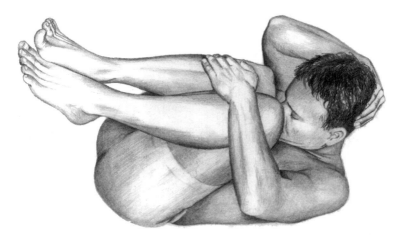

Note: In the exercise on page 89, we experienced a similar position as we curled up the body and embraced it. The exercise here is less active and should be done quite gently. The muscles are very relaxed after the initial phase.

We now let ourselves fall and imagine that we are weightlessly floating in water. Thousands of sensations rise up from the subconscious. It suffices to briefly perceive them and then let go of them—we do not have to concern ourselves with them any longer here.

C) Exercises Lying on the Stomach

C1
Grasping the Ankles and Pressing the Heels Against the Buttocks

Practicing flexibility

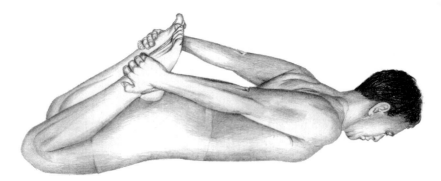

Note: This exercise is strenuous and also tortuous for people with very tense muscles. The entire body musculature is deliberately tensed while doing the exercise—up to the threshold of pain.

Tensing and relaxing follow the breathing rhythm—a good opportunity to consciously really experience the alternation between tension and relaxation. This exercise is an act of strength. People who can touch the buttocks with their ankles show that they are flexible.

It is usually necessary to make a number of attempts; at first, there are still thirty, then twenty, then just ten centimeters to success.

C2
Massaging the buttocks

Awakening the sensuality

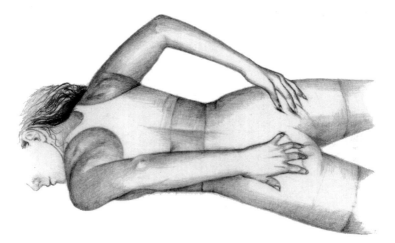

Note: In as far as it is possible, we attempt to do this exercise alone.

The gluteus muscles must bear our entire weight when we spend most of our day in a sitting position. The flow of blood becomes blocked. This is why we often feel the desire to get up and take a few steps.

We can easily relax the blocks by touching, kneading, and making these muscles come back to life again. By unloading not only our physical, but also our emotional weight on our bottom, we block an important erogenous zone. Just a few seconds of this massage is enough to take us from a feeling of heaviness and numbness to a feeling of elasticity and flexibility.

C3
Placing the Hands Around the Waist and Stroking the Kidneys

Opening up to friendship

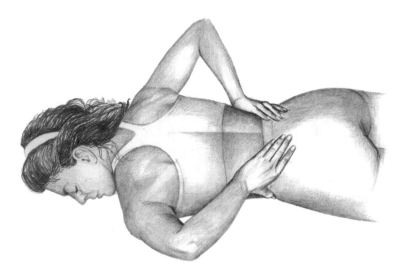

Note: In the symbolic language of the body, the kidneys represent our relationship to friendship and closeness.

This function of the kidneys is important for us: we feel it as soon as we put our "warm" hands on the kidney region and remain there a while—and rediscover an organ which is often forgotten.

C4
Placing the Head on the Elbows and
Press Against a Pillow

The liberating crying, the catharsis

Note: Those who can no longer cry, can no longer free their hearts of a burdensome heaviness.

The psyche unloads itself when we weep—and we feel ourselves to be freer afterwards. If there is no longer anything which we can still cry about, we simply start to imagine sad situations from our life—until our heart opens up. Then the tears begin to flow, the crying fit jolts us and we are liberated.

D) Erotic Exercises

D1
Consciously Imitating the Movements of Sexual Intercourse

Consciously experiencing the sexual exchange

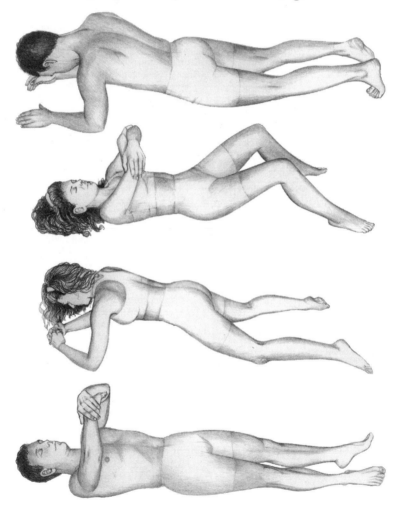

Note: When we move our muscles in the manner of sexual intercourse, we trigger a spectrum of associations. In our imagination, we can relive the sexual exchange. In this process, we observe ourselves as if from a distance, seeing everything very clearly, and can understand and project ourselves into the scene.

In order to understand our partner—which is an absolutely essential prerequisite for a true exchange—we also imitate his or her movements. The man then feels what the woman feels and also the other way around. With this, we intensify our understanding of each other.

While doing this exercise, we release testosterone and/or estrogen. This not only gives our sexual activities much strength, but can also provide us with important hormones when making an important occupational decision for which we need much energy and strength.

Whether we do two, several or all of the suggested exercises is not important.

In any case, it is good to take the balanced position after our exercises and relax even more as a result—and in our thoughts once again go through the exercises and let our perceptions run through our mind. Perhaps we will remember which trigger points we could feel most intensely.

s possible that the blood pressure has sunk somewhat because of the exercises. For this reason, we should get up slowly and strengthen ourselves through rhythmic tensing and relaxing.

It is particularly pleasant if we can take some time in order to return to our customary equilibrium. The best way to do this is to sleep for thirty minutes.

Deepening the Contact Between the "I" and the "You"

Exercises for Partners and Friends

Preliminary Remarks

The partner exercises are designed in such a manner that the roles are to be exchanged because it is important for each of the partners to also feel with the other.

In the following partner exercises, the "client"—indicated by "partner A"—is passive and relaxes physically and mentally as much as possible, working only with his or her powers of imagination.

The action part, the treatment, is carried out by the "therapist"—indicated by "partner B."

If the individual exercises are carried out in advance, this naturally intensifies the effect of the partner exercises.

In addition, it is very helpful to have an exchange of feelings and experiences after the exercises.

Summary of Partner Exercises

1

Partner B sits with straddled
legs at the feet of partner A,
grasps her ankles and "pulls"
her towards himself.

2

Partner B lifts and lowers
partner A's hips.

3

Partner B massages partner
A's chest according to his
wishes and instructions.

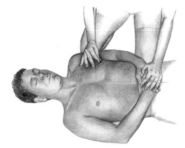

4

Partner B sits at partner A's
head with spread legs, holds
the hands under partner A's
nape and brings partner A's
head closer to partner B's
own genital organs.

5

Exercise for the woman:
She lies on her back, partner
B sits at her side and gently
puts a hand on the ovaries.

6

Exercise for the man:
He lies on his back, partner
B sits at her side and gently
puts a hand on the testicles.

7

Partner A lies on her back,
partner B lies and sits next
to her and puts the lower arm
on the line from the pubic
bone to the middle of the
chest.

8

Partner A and partner B lie
next to each other on the
side, faces turned to each
other, whereby A puts his
thighs between B's legs so
that the chakra lines cover
each other.

9

Partner A massages B's buttock muscles and then the other way around.

10

Partner A lies on the stomach, partner B sits at his side and massages the spinal column from the first cervical vertebra to the coccyx

11

Partner B's fingers feel the first cervical vertebra and gently press it downwards, at the same time the index finger of the other hand presses on the eminence of the coccyx and "presses" the spinal column together.

12

Partner A sits with opened legs, partner B leans against his chest, tenderly embraced.

Partner Exercises

1

Partner B sits at partner A's feet, grasps her ankles and pulls A's body closer in order to put the feet against B's ankles. Partner B's hand can grasp A's foot and massage the sole of the foot. Afterwards, the hand lightly strokes along the inner side of A's legs, from the ankles to the groin.

There are many trigger points on the inner side of the thigh (in the man, it includes the point from which the nerves lead to the testicles). Partner B now tries to localize these points on the basis of A's directions in order to alternately stroke them with light to very light pressure.

An interesting variation: Bring the soles of A's feet in contact with B's chest and observe the perceptions.

Coming closer to each other, making contact

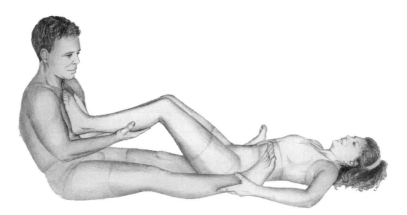

Note: The sole of the foot against the breast just seems to be a peculiar type of encounter. Two different body zones actually meet here and evoke sensations which have been unknown up to now, thereby increasing the intimacy. Body and soul come closer to each other.

2

Partner B sits at A's side and grasps the pelvis with his hands, lifting and lowering it a number of times in an agreeable rhythm. Together they do the exercise while A lies on the stomach and on the back. The yang movement is a strong rhythmic pressing the pelvis downwards. The yin movement is a gentle lifting of the pelvis, an opening up to the partner.

Yin and yang, feeling giving and taking

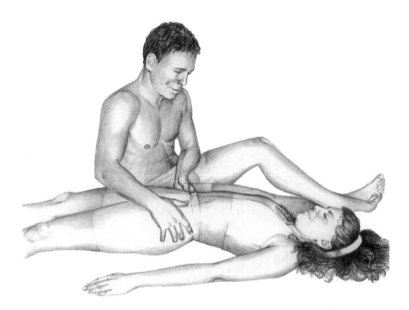

Note: These are apparently movements typical of the sexual exchange. However, the exercise is consciously carried out without erotic thoughts. The man feels what the woman feels and vice versa. Although the sensations are strong, we experience everything only on a purely mental level.

3

Partner B massages A's chest and listens to the latter's wishes and directions. We should remember that the masculine chest becomes very sensitive when touched lightly. The "massage" can also consist of putting the hand on the chest in the form of a cup.

The desire for being touched on the nipple naturally requires a special openness for this sensation, as well as a gentle approach by the person giving the treatment.

The simple way to learn to understand the partner's wishes

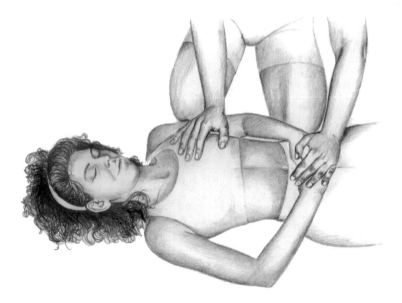

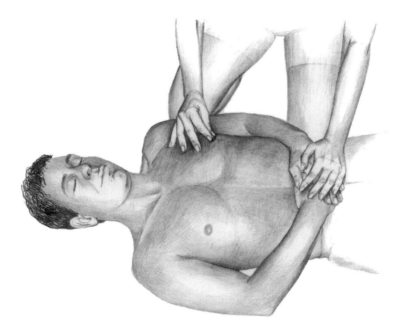

Note: This exercise is recommended for people who are not familiar with each other's bodies. In this situation, it is usually easier to express personal wishes. However, above all it is easier to really express one wish or another impassively—which is not possible during an intimate sexual exchange.

4

Partner B sits with spread legs in front of partner A's head, puts his hands under partner A's nape and puts it either on or close to his own genital organs.

How close the two people should come while doing this depends upon their degree of intimacy. There can be skin contact, but there does not have to be any at all.

The task of this exercise is to stay in this position for some minutes without awakening erotic feelings. At the same time, there is an intensive exchange of energy.

The "I" and the "you" merge with each other

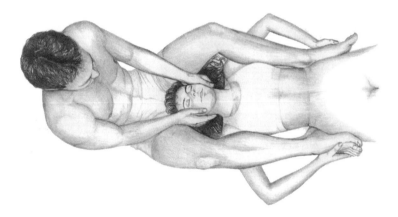

Note: This is certainly a very intimate position, but our emotional attitude is what is important. We also avoid erotic feelings here, which then creates an intensive, tangible exchange of energy.

5

This is an exercise especially for the woman: She lies on her back, B sits at her side and gently puts a hand on the ovaries, whereby the edge of the hand presses on the pubic bone. The ovaries make themselves noticeable in that they radiate heat.

Releasing vital hormones in the woman

Note: The sensations become stronger when both hands, that of partner B and partner A's hand, are put on top of each other. The flow of blood to this zone, of central significance for the woman, is very strong. Partner B remains passive and can observe how the expression of femininity becomes more intense, particularly in her face.

6

This is an exercise especially for the man: He lies on his back, partner B sits at his side and gently puts a hand on the testicles. Putting partner A's and B's hands on top of each other can increase the effect. Afterwards, partner B's fingers exercise a very gentle pressure on the inner side of the thigh.

Releasing vital hormones in the man

Note: The influence created by the release of hormones becomes apparent here. The exercise can have an erotic effect—but it doesn't have to. The vital hormones are just as important for the powerful ability to visualize as they are essential for one's own achievement.

7

Partner A lies on the back, partner B lies or sits next to her in a position which allows the underarm to lie on the entire line from the pubic bone to the upper center of the chest.

Going from a state of tension to relaxation to achieve inner peace

Note: The resting arm touches the entire energy centers along the midline from the stomach to the chest. Partner A has a feeling of increasing fatigue, which is why the exercise is ideal—particularly when there is a loving partner next to you who is willing to help—for difficulties in falling asleep.

8

The partners A and B lie on their sides next to each other, look in each other's eyes and talk together while their hands are stroking each other. At the same time, partner A puts her thighs between partner B's legs so that the energy lines cover each other.

When doing this, an extremely intensive body contact occurs which is an ideal starting point for a pleasant dialog equally intensified by the body and the mind. In this position, there is an exchange between the partners' chakras—root, heart and forehead chakras are facing each other.

Intimacy: Contact between mind (dialog), "heart" (feelings) and sexual zones (instincts)

Note: In this manner, the dialog is deepened through a very intensive physical contact. Feelings are communicated through contact of the hands and the instinctual sensations through the thighs. The partners are then connected at the same time with the three layers of the personality—mind, feelings and instincts.

9

Strong massage of the buttock muscles.

Reviving the sensuality

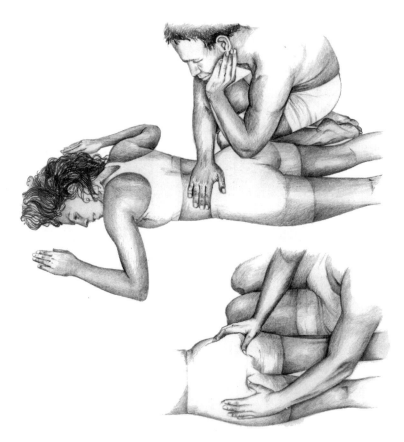

Note: Through no other body part is it possible to experience the other person in such a sensual, pleasant way. A non-erotic body contact is created.

We should not forget the hidden sensuality of this zone.

10

Partner A lies on the stomach. Partner B sits at his or her side and massages the spinal column from the first cervical vertebra to the coccyx and back up again. The treatment is repeated by changing from the initial kneading movement to a light stroking movement (as soon as the spinal column is "sensitized").

Feeling the spinal column as the support of the personality ...

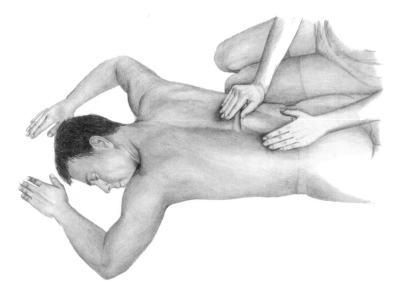

Note: Through this massage, we awaken the perceptions of the spinal column as the support of one's own personality. The emotional effect is sustained and the posture improved.

11

Partner B's fingers feel partner A's first cervical vertebra and gently press it downwards; at the same time, the index finger of the other hand (caution: although this can be very painful for some people, the pressure should be as strong as possible) presses on the lower eminence of the coccyx in the direction of the head.

... and your self is in my hands

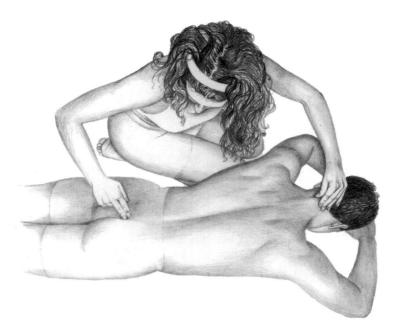

Note: It is easy to identify with one's own spinal column when it is touched and (com)pressed on the two extreme points so that it becomes one piece which is elastic, flexible and soft. The flow of energy from the body to the mind is felt.

In this exchange of giving support and leaning on someone, both partners are active and passive at the same time, melting into each other. The soft front side of the chest comes into contact with the hard back.

Leaning on the trunk

Note: There are very few other possibilities in which we can form a contact of this type. It is an interplay of devotion and giving support, embracing and giving protection at the same time.

Feeling the Contact to the Human Environment

An Exercise Which Can Be Done in a Group

Introduction

One persons lies on the back and the others group in a circle around him or her. The exercise, in which greatly different variations are possible, now begins.

The person sitting at the end where the head is can massage the skin of the head or take position 6 or 4 or one of the other positions.

Those sitting at the sides have just as many possibilities of touching. They treat the area from the groin to the knee.

The person sitting at the end where the feet are treats the ankles and soles of the feet. The people sitting across from each other at the sides carry out a variation of what the person across from them is doing.

After about 10 minutes, the person sitting at the end where the head is changes places with the one lying in the middle; the others shift in a clockwise manner.

The Exercise

The "self" at the center of attention, an experience of intimate closeness

Note: The group exercise is particularly valuable for all contact problems. In greatly varying group situations, this is a good opportunity to break through the wall of distance in interpersonal relationships.

We soon learn that:
- above all, making contact is pleasant
- it is even more pleasant "to be pampered" by many hands which are warm and loving. This results in a feeling of being understood and of personal esteem.
- other people are interested in you—as the center of an emotional world
- skin contact is the most intensive communication
- well-being is truly close at hand.

In Closing:
Change is Growth

Even if our "self" is a stable unity of body and mind to a certain extent, we should not forget that change is necessary for its development—according to the tasks that life gives us and the objectives which we set for ourselves.

This is a perpetual process, as natural as growth in nature. Plants or trees do not block their growth for fear of the thunderstorm which will certainly arise and for fear of lightning and thunder. They simple strengthen their roots and keep themselves flexible.

In order to consciously perceive this inevitable changeability within us, we should make a very simple inventory of our current condition—as a starting point, so to speak. To this we should gradually add notes in relation to our experiences during and after the exercise—they do not need to be long, but clearly and precisely formulated. They will help us perceive and follow our personal development and growth. All other goals are small changes necessary for us to reach our great goal. Examples of these small changes are our posture, our attitude, our interpersonal relationships and the decisions required for the arrangement of the course of our day.

The exercises can help us to strengthen the individual weak spots or change things that bother us.

If we think that we are lacking in the area of posture, which means the backbone, we will have the greatest success by working on the spinal column and soon feel it to be strong once again. A bent posture which leans forwards becomes upright (see the exercises on pages 108, 132 and 133).

The exercises on pages 92, 93, 94 and 114, 122, 127 and 128 give us new energy and vitality.

In the case of despondency or depression, the exercises on pages 121, 122, 127, 128 and 134 are particularly helpful, in addition to the group exercise on page 136.

However, the deeper sense of the exercises is achieving the turning point in one's own development in order to then take the path to the whole "self." This happens when we find the way from the extremes of mind-body to the actual center, when our "self" has found the free space for its own development, perceives itself and takes its own path.

We free ourselves by learning to listen to our own sensations and start to feel our "heart" as the emotional center.

What we would like to give you with this book could also be termed "esoteric," which means something like "the trusting, personal-intimate communication of *the experienced one* with students and friends." You can now absorb, try out and enjoy our thoughts and knowledge—thanks to the constructive and inspiring support of our publisher Monika Jünemann—with this book for yourself at home.

I welcome and am open for an exchange of experience and therefore sincerely invite you to tell me of your experiences with the exercises (see address on page 141). In as far as possible, I will personally answer you, and certainly with further comments in addition to the suggestions made here. The exercises—as introduced here—have also been created with the support of the mental exchange with my friends and work groups.

Here are just a few examples that show how extensive the changes are which the exercises can trigger.

Anna, 32, was inhibited in the entire area of emotional expression: "... as the fingers touched the solar plexus, it felt as if they were penetrating into me, and at the same time, I felt how the chest and abdominal muscles relaxed. I had the feeling of opening up widely, of finally being able to be open ..." (Exercises on pages 97 and 100).

Karolina, 26, inhibited in the pre-orgasmic sensations: "...the sensations of the exercises on pages 126, 127 and 130, which are simply considered to be non-erotic, led me into a state of thought-free relaxation, like I could never achieve during sexual exchange and which were almost an emotional orgasm."

Michael, 27: "It is my goal to become a music conductor, but ... I am a fearful person, my vegetative nervous system is extremely sensitive. I constantly pay attention to my heart. Every peculiar perception excites me and brings me into a state of alarm ... soon afterwards, my heart begins to race ... and I ask myself whether this is an indication from my body that I should give up this too ambitious of an objective. After two weeks he discovered: "... These days I do the relaxation exercises on pages 66, 99, 124 and 129 either alone or let Theresa help me. I haven't found the full contact to my body yet, but the exercises are both pleasant and relaxing. I feel it to be most pleasant when Theresa massages the soles of my feel and grasps the ankles. In this way, I find the contact to the earth and have a feeling of self-confidence. ... now I am also starting to see my heart, which I have feared up to now, as a friend, my fear is diminishing and I can perceive it without getting into a state of panic."

About the Author and His Work

Franz Benedikter is a Doctor of Philosophy. He has a practice close to Rome, Italy. This is where he imparts to his clients how they can restore and intensify the positive contact to their bodies through touch therapy with endogenous induction and thereby create the best possible hormonal basis for a healthy, happy and liberated life.

His practice rooms are conceived in such a manner as to support the "therapy" in the best way possible. The room for verbal client-centered therapy is a dome-shaped building, the spatial aura of which positively supports an opening of the mind. The exercise room for individuals, pairs and groups radiates warmth, softness and "grounding." This room supports the perception of one's own physical space and the "grounding" of the people who work on themselves here.

"Tempietto" discussion room

"Tempio" exercise room

Addresses

Those who would like to get in contact with Franz Benedikter or are interested in training in "Endogenous Induction" can write the german publisher with an addressed and stamped return envelope (please enclose an international reply coupon). They would be glad to pass your personal letters on to the author. Please do not forget the catchword.

Windpferd Verlag
Catchword: "The Secrets of Loving Touch"
Postfach
D-87648, Germany

Literature and Music Tips

Baginski, Bodo J. und Sharamon, Shalila, *The Chakra Handbook*, Lotus Light, Twin Lakes, WI, 1989. This is an excellent work about the functions of the seven main chakras with many exercises, classification tables and thought-provoking ideas.

Diamond, Dr. John, **"Life Energy"**, self-published. This book discusses kinesiology in reference to acupuncture meridians, reflex zones for testing and treatment and shows an interesting correlation to biochemistry.

Horan, Paula, *Empowerment Through Reiki*, Lotus Light Publications, Twin Lakes, 1994. This is an important book about Reiki. Besides that it describes many of the ideas of holistic medicine.

Lübeck, Walter, *The Complete Reiki Handbook*, Lotus Light Publications, Wilmot, WI, 1995. A detailed introduction oriented toward the practice of Reiki healing. There is an extensive ABC of special positions, as well as advice concerning Reiki and medication, Reiki meditation, work with precious stones and aromatherapy.

Mann/Short, *The Body of Light*, Charles E. Tuttle, Boston. This is a very interesting introduction to various ways in which the great spiritual traditions look at the inner energy system.

Music to create a peaceful and lovely spirit

Merlin's Magic, *Reiki Musik* and *Reiki – The Light Touch*. Blissful sounds when ever you want to feel wonderful. Perfect for body work, meditation, relaxation and, of course Reiki healing sessions. (Inner Worlds Music, Boulder)

Dancer's Dream, *Dance of Dreams* is a gently sensuous Tantra-lovers delight filled with melodies brimming with passion and spiritual ecstasy! Wonderful for moving meditation, trance dance or a positive energy tune-up. (Inner Worlds Music, Boulder)

Ralf Bach, *Desire for Love, More Desire for Love* and *Angel Kisses*. Perhaps the most romantic music ever recorded. These lushly orchestrated instrumental melodies are perfect for intimate occasions, quiet times or wehen you want to escape into a beautiful world of blissful daydreams – accented with soothing sounds of nature. (Inner Worlds Music, Boulder)

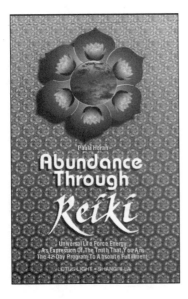

Paula Horan

Abundance Through Reiki

**Universal Life Force Energy
As Expression of the Truth That
You Are
The 42-Day Program to Absolute
Fulfillment**

Abundance Through Reiki is a powerful, poetic evocation of true self and universal life force energy. Its emphasis is a program of 42 steps from Core Self to Core Abundance, creating inner and outer richness. A detailed presentation in the form of two 21-day abundance plans takes you on an exploration of belief patterns that keep you from experiencing everything you need or desire.

Further topics are Reiki and abundance, abundance of health, love, friendship, knowledge, and experience. The book promotes your own natural ability to experience freedom, creativity, and authenticity.

160 pages, $14.94
ISBN 0-914955-25-X

Shalila Sharamon · Bodo J. Baginski

The Healing Power of Grapefruit Seed

The Practical Handbook for Using Grapefruit Seed Extract to Heal Infections, Allergies, and Much More. One of the Most Effective New Healing Remedies

Latest scientific studies show that plant extract from grapefruit seeds has a large range of effects and applications for both internal and external use in preventative health care, therapy, cosmetics, and baby care. Based on international research, two bestselling authors have compiled sensational therapy successes and areas of application for this biological broad-spectrum therapeutic agent, antibiotic, antimycotic and antiparasitic, preservative, and hygienic agent of the future. In addition to scientific proof, this practice-oriented book includes proper dosages and procedures.

160 pages, $ 12.95
ISBN 0-914955-27-6

Rodolphe Balz

The Healing Power of Essential Oils

Fragrance Secrets for Everyday Use. This handbook is a compact reference work on the effects and applications of 248 essential oils for health, fitness, and well-being

Fifteen years of organic cultivation of spice plants and healing herbs in the French Provence have provided Rodolphe Balz with extensive knowledge about essential oils, how they work, and how to use them.

The heart of *The Healing Power of Essential Oils* is an essenial-oil index describing their properties, followed by a comprehensive therapeutic index for putting them to practical use. Further topics of this indispensible aromatherapy handbook are distillation processes, concentrations, chemotypes, quality and quality control, toxicity, self-medication, and the aromatogram.

208 pages, $ 14.95
ISBN 0-941524-89-2

Walter Lübeck

Reiki – Way of the Heart

**The Reiki Path of Initiation
A Wonderful Method for Inner Development and Holistic Healing**

Reiki – Way of the Heart is for everyone interested in the opportunities and experiences offered by this very popular esoteric path of perception, based on easily learned exercises conveyed by a Reiki Master to students in three degrees.

If you practice Reiki, the use of universal life energy to heal oneself and others, you will have the possibility of receiving direct knowledge about your personal development, health, and transformation.

Walter Lübeck also presents a good survey of various Reiki schools and shows how Reiki can be applied successfully in many areas of life.

208 pages, $ 14.95
ISBN 0-941524-91-4